AF691126

# FACILITATION AND DEBRIEFING IN AVIATION TRAINING AND OPERATIONS

# Facilitation and Debriefing in Aviation Training and Operations

*Edited by*
R. KEY DISMUKES
*NASA Ames Research Center*
GUY M. SMITH
*Ypsilon Associates*

**Ashgate**

Aldershot • Burlington USA • Singapore • Sydney

© R. Key Dismukes and Guy M. Smith 2000

All rights reserved. No part of this publication may be reproduced, stored in a retrieval system, or transmitted in any form or by any means, electronic, mechanical, photocopying, recording or otherwise without the prior permission of the publisher.

Published by
Ashgate Publishing Ltd
Gower House
Croft Road
Aldershot
Hants GU11 3HR
England

Ashgate Publishing Company
131 Main Street
Burlington
Vermont 05401
USA

Ashgate website: http://www.ashgate.com

**British Library Cataloguing in Publication Data**
Facilitation and debriefing in aviation training and operations. -
(Studies in aviation psychology and human factors)
1.Airplanes - Piloting - Study and teaching 2.Air pilots -
Training of 3.Group facilitation
I.Dismukes, Key II.Smith, Guy M.
629.1'3252'071

**Library of Congress Control Number:** 00-133536

ISBN 0 7546 1164 7
Printed and bound in Great Britain by MPG Books Ltd, Bodmin, Cornwall

# Contents

# List of Contributors

**R. Key Dismukes, Ph. D.**
NASA Ames Research Center
MS 262-4
Moffett Field, CA 94035-1000, USA
e-mail: kdismukes@mail.arc.nasa.gov

**Kimberly K. Jobe**
San Jose State University Foundation/
NASA Ames Research Center
MS 262-4
Moffett Field, CA 94035-1000, USA
e-mail: kjobe@mail.arc.nasa.gov

**Captain Mark B. Klair**
Critical Communications
8956 Morning Grove Cove
Cordova, TN 38018, USA
e-mail: mklair@aol.com

**Lori K. McDonnell**
San Jose State University Foundation/
NASA Ames Research Center
MS 262-4
Moffett Field, CA 94035-1000, USA
e-mail: lmcdonnell@mail.arc.nasa.gov

**Guy M. Smith, Ph. D.**
Ypsilon Associates
Lakeville, MN 55044, USA
e-mail: ypsilon@earthlink.net

**Captain Mimi Tompkins**
Aloha Airlines
PO Box 31173
Honolulu, HI 96820, USA
e-mail: mltompkins@compuserve.com

**Captain Peter J. "Pete" Wolfe**
Southwest Airlines, Flight Operations Human Factors
P.O. Box P.O. Box 36611
Attn: HDQ - 8FO
Dallas, TX 75235, USA
e-mail: pjwolfe@earthlink.net

# Acknowledgements

The editors wish to thank Lori McDonnell and Kimberly Jobe for their work in the preparation of this volume and their review of the various drafts of the contributed chapters. We would also like to acknowledge Amber Laurel's and Krista Barber's assistance with the manuscript preparation. The chapter authors have taken considerable time from their personal lives to share their experiencces with facilitation in the aviation community so that this volume could become a reality. Eleana Edens, program manager of the Federal Aviation Administration's Office of the Chief Scientist and Technical Advisor for Human Factors, encouraged and funded the research and training manual described in chapters 2 and 3. Finally, we would like to thank John Hindley and Anne Keirby at Ashgate Publishing Group for their support in the production of this book.

# 1 What Is Facilitation and Why Use It?

R. KEY DISMUKES, LORI K. MCDONNELL, KIMBERLY K. JOBE, AND GUY M. SMITH

Facilitation, as the term is used in this book, refers to processes by which one member of a group operates to help the group analyze issues, learn from experience, and work as a team to draw conclusions. In aviation settings, the facilitator is typically an instructor, however facilitation requires the instructor to work in ways quite different from those of traditional instruction. The traditional roles of the aviation instructor have been to convey information, typically to trainees who are less knowledgeable than the instructor, and to evaluate trainee performance. Facilitation, in contrast, focuses on assisting trainees to learn and evaluate through their own inquiry. Facilitation is especially appropriate when trainees already possess substantial expertise.

Facilitation in aviation training began with the early developers of Crew Resource Management (CRM) and Line Oriented Flight Training (LOFT), who recommended that instructors facilitate crew debriefings of LOFT rather than lecture the crew on what they did right or wrong. Over the past twenty years this concept has become widely accepted: unfortunately, until recently detailed materials and methods specific to aviation have not been available to train instructors in the demanding skills of facilitation. Several individuals and organizations have experimented with using facilitation in other aspects of aviation training and operations, however until now little of this work has been published in a widely available form. We intend this book to show the innovative ways facilitation is being used in aviation and to provide a detailed account of methods of facilitation that will enable aviation professionals to apply these methods in their work.

In this chapter, we outline the basic concepts of facilitation and trace its origins from the fields of education and humanistic psychology. We compare the advantages and disadvantages of facilitation with traditional methods of instruction and discuss the circumstances in which each is best used, and we recommend an approach to training facilitators.

Chapter 2 describes a research study of facilitation in Line Oriented Flight Training (LOFT) debriefings, and chapter 3 gives detailed guidance for using facilitation techniques in LOFT debriefings. Chapter 4 identifies other applications of facilitation in aviation training. The concept of crew-centered – as opposed to instructor-centered – debriefing can be extended beneficially to other aspects of flight operations. Chapters 5, 6, and 7, respectively, discuss debriefing of normal line operations, critical line incidents in which crews have gotten into some sort of trouble, and traumatic line incidents and accidents.

## The Concept of Student-Centered Learning

The progressive movement in education advocated a shift in emphasis from the teacher to the student (Elias & Merriam, 1980). John Dewey, the chief advocate of the progressive approach to education, argued in his earliest writings that the role of the teacher is to provide a setting that is conducive to learning (Dewey, 1916; Elias & Merriam, 1980). According to Dewey, learning is something students do for themselves, so once the teacher has provided the conditions that stimulate learning, the rest lies with the learner. In later writings Dewey described teachers as leaders of group activities, who must survey the needs and capacities of individual learners and create the conditions that meet these needs (1938). Dewey also suggested that teachers should share insights that come from their own experiences without imposing their own views on the learners.

Although Dewey was primarily concerned with education of children, his concept is even more applicable to adult learning. Seaman and Fellenz (1989, p. 5) state: 'Much of the research in teaching adults indicates that active participation by the learner and meaningfulness of content are two constant factors influencing the effectiveness of the teaching/learning process'. Zemke and Zemke (1981), reviewing studies of what adults like and dislike in the classroom and in meetings, concluded that adults dislike long lectures, that they learn best from discussions with their peers, and that their self-esteem is on the line because they tend to take things personally. Zemke and Zemke also concluded that it is critical to articulate and clarify all expectations up front and that new knowledge must be integrated with old through active participation.

Brookfield (1986) traces the idea that educators should function as facilitators of learning, rather than disseminators of knowledge, to the field of humanistic psychology, especially the work of Carl Rogers. According to Rogers (1969), there are two types of learning, divided along a continuum of meaning. The first type involves the mind only, as in memorizing facts that have no personal meaning for the individual. Rogers refers to this type of learning as 'from the neck up' – it does not engage the whole person because it does not involve feelings or personal meaning. In contrast is experiential learning, which is significant, meaningful and self-initiated. Even when an outside source stimulates learning, the sense of discovery and comprehension comes from within. When we discover something meaningful for ourselves, we incorporate our thoughts and feelings by being personally involved in the learning event. This personal involvement results in learning that is more pervasive; it effects the behavior, attitudes, and possibly even the personality, of the learner.

Based on his own experiences, reports of experiences of other facilitators of learning and relevant research, Rogers (1969, pp. 162-163) enunciated several principles of facilitation of learning. Three of these are particularly relevant to facilitation in the aviation setting:

- Learning is facilitated when the student participates responsibly in the learning process. Students learn substantially more by participating actively rather than passively.
- Self-initiated learning that involves the whole person of the learner – feelings as well as intellect – is the most lasting and pervasive. This 'gut-level' type of learning arises from activities such as developing one's own ideas and learning difficult skills.
- Independence, creativity, and self-reliance are all facilitated when self-criticism and self-evaluation are primary and evaluation by others secondary. It is through the evaluation of our own behaviors that we learn from our mistakes and our successes so we can become self-reliant.

Closely related to Rogers' ideas is the concept of active learning. Bonwell and Eison (1991, p. 2) list the general characteristics of active learning:

- Students are involved in more than listening.
- Less emphasis is placed on transmitting information and more on developing students' skills.
- Students are involved in higher-order thinking (analysis, synthesis, evaluation).
- Students are engaged in activities (e.g., reading, discussing, writing).
- Greater emphasis is placed on students' exploration of their own attitudes and values.

The objective is to increase learning by elevating the level of student engagement with the material to be learned. Active learning promotes increased student involvement in the learning process and it supports instructional strategies such as discussion-leading and skillful questioning techniques to engage the learners in personal exploration of the subject matter. The active learning approach seeks to dispel the 'Container-Dispenser Model' of instruction in which knowledge is a substance, the source of power; instructors are containers, filled with content, material and facts; and students are the vessels, wanting to be filled up (Pollio, 1987).

The concept of student-centered learning is quite consistent with modern research in cognitive psychology. Active participation requires students to process information more deeply than does merely listening passively (Slamecka & Graf, 1978). Deeper processing elaborates information in long-term memory structures in a way that enables better retrieval when the information is needed (Baddeley, 1990). Perhaps most important, the student-centered approach allows the learner to incorporate new information into the established framework of the learner's existing memory structures, which facilitates retention of the information and application to situations that may occur long after the initial learning (Anderson, 1990).

The objective of most aviation training goes beyond having the trainee simply acquire information. Typically it is crucial that the trainee incorporate that information and apply it in operations with a high level of skill; thus mastery of the information and its implications is required. In his Taxonomy of Educational Objectives, B. S. Bloom (1956) argued that there are six levels of mastery, arranged hierarchically by the level of mental complexity involved (see also Downing, 1995). The six levels, from least to most demanding, are knowledge, comprehension, application, analysis, synthesis, and evaluation. (Note that in the terminology of cognitive psychology what Bloom called knowledge would be termed 'information'.) Most aviation training requires all six levels of mastery. For example, in recurrent LOFT, crews are assumed to already know and comprehend the basic principles of CRM; the LOFT simulation provides an opportunity to apply those principles. The debriefing following the LOFT should provide crews the opportunity to achieve the highest levels of mastery: analysis, synthesis, and evaluation. Crews analyze what happened in their LOFT exercise and explore the relationships among events, crew actions, and outcome. From this analysis crews can synthesize their own ideas of how to deal with situations in line operations, and they can learn to evaluate their own performance meaningfully. Facilitation helps crews achieve these higher levels of mastery.

## The Role of the Facilitator

The previous section presented evidence that the student-centered approach offers advantages for aviation training; a later section of this chapter examines when to use the student-centered approach and when to use more traditional instruction methods. Taking a student-centered approach does not eliminate the need for an instructor, rather it shifts the instructor's role from dispensing information and judging student performance to facilitating student learning and self-evaluation. How does one facilitate another person's learning? To answer that question, we start by examining the origin of facilitation techniques.

Carl Rogers (1969) developed the humanist concept of the facilitator as a catalyst, one who uses skilled questioning techniques to help clients draw their own conclusions from their personal experiences and create their own prescription for change. Rogers often worked in self-help and group therapy settings: In these settings, according to Rogers, the facilitator plays a key role in establishing the initial mood or climate that will shape how the group will function. The facilitator helps elicit and clarify the purposes of the individual members and the group itself. Overall, the facilitator is a flexible resource to be used by the group, a counselor or advisor to the group.

Although Rogers worked in the field of therapy, his ideas transferred to facilitation of a relatively 'healthy' population in the business and management arena. The term 'facilitator' was first widely used in Quality Circles in the 1970s. Consultants developed techniques to help business organizations analyze and solve problems. Calling themselves facilitators, these consultants expressed the notion that their job was not to solve the organizations' problems themselves but to facilitate groups within the organizations who were addressing the problems. These business facilitators, who typically lacked the specific subject matter expertise of the group members, focused on the processes by which the group worked together. Their objectives were to help the group clarify its goals and remain focused on those goals, to enable all of the knowledge and skills of the individual group members to be brought to bear on the issue, and to prevent any individual from dominating the discussion or cutting off relevant ideas.

To accomplish these objectives, Robson and Beary (1995) recommended that facilitators use certain tools, especially:

- *Developing empathy.* Facilitators must be able to see things through the eyes of the client or customer in order to develop a trusting relationship and to work effectively with that person. Facilitators can use specific techniques to build rapport and demonstrate empathy. These techniques include leaning forward while the other person is talking, active listening, and mirroring (echoing the physical posture of the other person).

- *Listening carefully.* Facilitators must hear what is really said, and avoid 'selective perception', hearing just what they want or expect to be said. Facilitators can demonstrate they understand what others are saying by reflecting back in their own words their understanding of what the other person seems to be saying.
- *Eliciting information.* Techniques include asking open questions that cannot be answered with a simple 'yes' or 'no' and using non-verbal communication (e.g., nodding, leaning forward, and smiling encouragingly while the other person is speaking). Eliciting information also requires facilitators to restrict the amount of talking they do themselves, so that the majority of the discussion is contributed by other group members.
- *Positive confrontation.* The facilitator must call attention to muddled, distorted, or inappropriate thinking in a way that enables the group to confront the problems. This must be done in a way that does not provoke defensiveness or loss of ownership by group members, and this requires great skill and tact. Positive confrontation is best accomplished by giving feedback that is descriptive but not judgmental. To avoid eliciting defensiveness, the facilitator may find it useful to preface this feedback with statements such as 'I may be wrong but it sounds as if it might be useful to consider other interpretations'. If individuals persist in arguing that their perspective is the only correct one, the facilitator can point out that regardless of which view is factually correct, we are all affected by others' perceptions and thus it is to our advantage to attend to those perceptions.

Although these tools were developed for use in business settings very different from aviation training, they can be applied quite effectively to facilitation in aviation settings. Chapter 3 provides detailed examples of how techniques based on these tools can be used in an aviation setting.

## Facilitation in Aviation Training

Crew debriefing following the simulation is a critical part of LOFT. From its early development in the mid-1970s, LOFT philosophy encouraged instructors to act as moderators who helped crews critically analyze and assess their own performance. Reporting on a NASA/industry workshop of LOFT held in 1979, Lauber and Foushee (1981) wrote:

> ...the role that the instructor plays during the debriefing session is primarily that of moderator. Because there are no 'right' solutions to many LOFT problems, it is more important for the instructor to guide the debriefing session, so that the full range of potential approaches to the problem is explored, rather than to impose his or her ideas about how the problems should have been handled. Experience has shown that crews frequently debrief themselves. Self-criticism and self-examination are almost always present in these situations and in many cases they are much more effective than instructor criticism...the instructor should do everything possible to foster this sort of self-analysis while at the same time keep it at a constructive level...the instructor should avoid lectures about what is right and what is wrong.

Clearly, by the time of the 1979 workshop the airline industry was envisioning a non-traditional role for instructors leading LOFT debriefings; however the workshop proceedings do not discuss techniques instructors might use when acting as 'moderators'. It is not clear who first introduced the concepts and techniques of facilitation into LOFT debriefing. The UK Royal Air Force was using the term facilitation in training ground and flight instructors in these new concepts in 1984 (N. McCloud, personal communication, January 2000). Delta Airlines was training instructors in facilitation skills in 1989 (Byrnes & Black, 1993). Continental Airlines provided its instructors a guidebook dated 1992 that explicitly described the advantages of learner-centered training and described the kinds of facilitation techniques used in business settings (Continental Airlines, 1992).

Adapting facilitation to debriefing requires some modification of the process because the debriefing differs in several important respects from both business and therapy settings. In most business and therapy settings individuals participate in groups more or less voluntarily. In contrast, airline companies mandate that crews participate in LOFT debriefings. Authority and responsibility for meeting training objectives lie with the instructor, and the crews do not necessarily bring specific ideas of what they want out of the training. In contrast to facilitators in business settings, LOFT instructor/facilitators have deep subject matter expertise; furthermore from their vantage point in back of the simulator cab they observe aspects of technical and CRM performance that the crew might not observe themselves.

These differences, however, do not impose any insurmountable obstacle, and LOFT instructors can readily work within the general concepts of facilitation if they are adequately trained to do so. Airline pilots are highly professional; most take pride in their performance, seek out ways to hone it constantly, and enjoy helping other crewmembers to do the same. Thus the crew's motivation is consistent with the objectives of facilitation. Instructors can integrate their own expert observations into the crew's discussion in the debriefing by acting as a jumpseat member of the crew, whose expert observations and knowledge are highly relevant – as are those of all the crewmembers.

## Facilitation Versus Traditional Instruction

We do not want to give the impression that traditional instruction should be replaced across the board with facilitation in aviation training. Both facilitation and traditional instruction have advantages and disadvantages. The particular objectives of the training should determine which approach is taken.

Traditional instruction is oriented toward transfer of information from the instructor to students in situations in which the instructor knows substantially more about the subject than do the students. In situations appropriate for facilitation, the emphasis is on applying existing knowledge and gaining insight, although information is often exchanged among the group participants. In these situations group members may have as much or more knowledge and expertise in the domain being discussed as the instructor. Traditional instruction can transfer large amounts of information fairly rapidly, but facilitation is often a fairly slow process.

Because the flow of information in traditional instruction is predominantly from instructor to students, the instructor does most of the talking and keeps any discussion tightly focused on predetermined objectives. In facilitation the facilitator strives to help the group determine its own objectives and encourages other group members to do most of the talking.

Traditional instruction typically requires testing to ascertain whether students have correctly assimilated the information transmitted to them. Facilitation deals with situations in which there is usually no one right answer; in these situations the facilitator is concerned with helping the group perform its own evaluations and reach its own conclusions.

For lecturing, the most intense period for the instructor is preparation – working up detailed information in advance so that it can be delivered effectively. Facilitation generally does not require extensive preparation before each session, but it is intensely demanding on the facilitator to draw out quiet members of the group, stimulate thoughtful analysis, keep track of the discussion, and help the group keep on track – all without dominating the discussion.

With these differences in mind, it is easy to see that some kinds of training are much better suited to traditional methods. It would be silly to try to use facilitation to teach the use of the flight management system to pilots who have never used the system. Conversely the LOFT debriefing is well suited to facilitation because the crews have substantial experience in flight operations, and they need to use their existing technical and CRM knowledge to analyze what happened in the LOFT simulation, to evaluate what they did well and not so well, and draw conclusions that they can carry back to everyday line operations. Facilitation is also well suited to other aspects of aviation training and operations described in chapters 4 through 7.

**What Can Go Wrong?**

Most U.S. airlines that we observed in the study described in chapter 2 provide at most very modest training in facilitation for their instructors. This is unfortunate, since facilitation requires skills substantially different from those of traditional instruction. An instructor minimally trained to facilitate LOFT debriefings – perhaps never even having seen a debriefing facilitated effectively – may find that crews do not respond as desired and conclude that facilitation is a poor method. Indeed in this situation the crews learn little.

Like the instructor/facilitator, the crewmembers are expected to operate differently in LOFT debriefing than they have in traditional training. If they have not previously participated in an well-facilitated debriefing, and if the instructor/facilitator does not provide a clear and convincing rationale for the crew-centered approach to debriefing, the crew may not grasp how they should participate.

A less obvious obstacle to facilitation arises because LOFT instructors must wear two hats. In most U.S. airline training programs the visit to the training center for LOFT is combined with recurrent technical training and proficiency checking. During the visit the same instructor may be providing traditional instruction and evaluating the crew's performance on maneuvers on which they will be tested later in the visit. In many cases the crews practice maneuvers ('batting practice') in the simulator between the LOFT and the LOFT debriefing (this problematic intermixing is discussed in chapter 2). Crews may find it hard to fully accept that the instructor's role in the LOFT debriefing is quite different from the rest of the visit, and they may at first be reluctant to expose themselves to the instructor by discussing problematic aspects of their performance. Furthermore, it is not entirely easy for the instructor to jump back and forth between two roles with such different objectives, and thus the instructor may unintentionally slip back into the more familiar role of lecturing, especially when he or she has noted shortcomings in the crew's performance in the simulator.

## Training Instructors to Facilitate

Facilitation requires techniques and skills quite different from those of traditional instruction. Most of the LOFT instructors interviewed in the study in chapter 2 reported they felt inadequately trained in facilitation. An hour or two of training and possession of a casually prepared company booklet on principles of facilitation are far from adequate preparation for becoming a facilitator. Little has been published to guide training of airline facilitators; what is available is often informally prepared and sketchy. We outline here our concept of the basic ingredients that should go into training LOFT instructor/facilitators. This training could also generalize to facilitation of other types of aviation training.

Initial classroom training should introduce instructors to the principles of facilitation and provide specific techniques that can be used. This chapter and chapter 3 give examples of the material that should be covered. Instructor/facilitator trainees should then be given ample opportunity to practice these principles and techniques, using their classmates as mock crews reenacting a simulated flight. Instructors are often quite good at mimicking typical crew behaviors and attitudes in debriefing. Each trainee should have several opportunities to facilitate a mock debriefing and encounter several of the most common problems that sometimes occur in debriefings. Among these problems are crewmembers who are reluctant to speak up, crewmembers that tend to be domineering, crews that fall into arguments, and crews that drift off into minute discussion of purely technical issues.

After each mock debriefing, the crew actors can give the facilitator trainee telling feedback on their experience of his or her efforts. This feedback is especially effective because those giving it are peers of the trainee. The facilitator trainees need specific criteria with which to evaluate their own performance and that of their classmates. We have found that his can be accomplished by giving he trainees the Debriefing Assessment Battery (DAB) described in chapter 2. Trainees report that the DAB helps them identify concrete goals for facilitation.

In order to conduct these mock debriefings realistically, the facilitator and crew actors must have a specific flight scenario in mind. Videos re-creating line incidents or accident re-creations can provide this background very nicely. After viewing the video re-creating a problematic event, the trainees playing the crew know what behaviors and attitudes to mimic, and the facilitator trainee knows what aspects of performance require examination.

After the classroom training, new facilitators should be assigned to a mentor, an instructor with substantial experience in facilitation and who is recognized to be effective at it. The new facilitator should have a chance to observe the mentor facilitate several debriefings and discuss those debriefings afterward with the mentor. Then the mentor should observe the new facilitator run several actual debriefings and give private feedback afterward.

It is helpful for LOFT instructor/facilitators to meet at least once a year to discuss their experiences in debriefing, to share problems encountered and brainstorm on techniques for getting around those problems. Facilitation techniques should not be considered carved in stone; once the basic principles are understood, facilitators can be quite creative in devising techniques to address the specific needs of the crews with whom they work.

We do understand that in the airline training business time is money, and that the facilitation training outlined here requires more time than most airlines have devoted historically. Nevertheless, few things can have more impact on the efficiency and safety of flight operations than enabling crews to analyze and improve their own performance. That is the goal of facilitation.

## References

Anderson, J. R. (1990). *Cognitive psychology and its implications.* New York: W. H. Freeman and Co.

Baddeley, A. (1990). *Human memory: Theory and practice*. Massachusetts: Simon & Schuster, Inc.

Bloom, B. S. (Ed.). (1956). *Taxonomy of educational objectives: Handbook 1: Cognitive domain.* New York: Longman.

Bonwell, C. C., & Eison, J. A. (1991). *Active learning: Creating excitement in the classroom*. ASHE-ERIC Higher Education Report No. 1. Washington, D.C.: The George Washington University, School of Education and Human Development.

Brookfield, S. D. (1986). *Understanding and facilitating adult learning.* San Francisco: Jossey-Bass.

Brookfield, S. D. (1989). Facilitating adult learning. In S. B. Merriam & P. M. Cunningham (Eds.), *Handbook of adult and continuing education* (pp. 201-210).

San Francisco: Jossey-Bass.

Byrnes, R. E., & Black, R. (1993). Developing and implementing CRM programs: The Delta experience. In E. L. Wiener, R. L. Helmreich, & B. G. Kanki (Eds.), *Cockpit resource management* (pp. 421-443). San Diego: Academic Press.

Continental Airlines, Flight Operations Human Factors Group. (1992, January). *LOFT facilitation techniques: Practical strategies and techniques for LOFT facilitators*. Unpublished manuscript.

Dewey, J. (1916). *Democracy and education*. New York: Free Press.

Dewey, J. (1938). *Experience and Education*. New York: Macmillan.

Downing, J. P. (1995). *Teaching for the whole brain: A practical guide for integrating creativity into learning*. Unpublished manuscript.

Elias, J. L., & Merriam, S. (1980). *Philosophical foundations of adult education*. Malabar, FL: Robert E. Krieger Publishing Co.

Lauber, J. K., & Foushee, H. C. (1981). *Guidelines for line-oriented flight training*. (NASA Conference Publication 2184). Moffett Field, CA: NASA-Ames Research Center.

Pollio, H. R. (1987). Practical poetry: Metaphoric thinking in science, art, literature, and nearly everywhere else. *Teaching Learning Issues*, No. 60. Knoxville: University of Tennessee, Learning Research Center.

Robson, M., & Beary, C. (1995). *Facilitating*. Aldershot, England: Gower Publishing Company Ltd.

Rogers, C. R. (1969). *Freedom to learn*. Columbus, OH: Charles E. Merrill.

Seaman, D. F., & Fellenz, R. A. (1989). *Effective strategies for teaching adults*. Columbus, OH: Merrill.

Slamecka, N. J. & Graf, P. (1978). The generation effect: Delineation of a phenomenon. *Journal of Experimental Psychology: Human Learning and Memory*, *4*(6), 592-604.

Zemke, R., & Zemke, S. (1981). 30 things we know for sure about adult learning. In P. G. Jones (Ed.) (1982), *Adult learning in your classroom: The best of training magazine's strategies and techniques for managers and trainers*. Minneapolis, MN: Lakewood Books.

# 2 Facilitating LOFT Debriefings: A Critical Analysis

R. KEY DISMUKES, KIMBERLY K. JOBE, AND LORI K. MCDONNELL

## Introduction

The airline industry uses Line Oriented Flight Training (LOFT) extensively to provide crews opportunity to practice Crew Resource Management (CRM) skills in realistic and challenging full-mission flight simulations. How much crews learn in LOFT and take back to their flying on the line depends in part on the effectiveness of the debriefing that follows the simulation (Helmreich & Foushee, 1993). The LOFT simulation is a busy, intense experience, and thoughtful discussion afterward is necessary for the crew to sort out and interpret what happened and why.

The concept of facilitated debriefings has been part of LOFT since its early inception (Lauber & Foushee, 1981; also see chapter 1, this volume). The facilitation approach contrasts markedly with most other aspects of flight training in which instructors critique the performance of pilots. The rationale for having instructors facilitate crew self-critique in LOFT debriefing is that crews can learn and remember much more when they participate actively and make their own analyses than when they listen passively to the instructor (Duval & Wicklund, 1972; Smith, 1994). Active participation requires crewmembers to process information more deeply than does listening passively (Slamecka & Graf, 1978), and deeper processing allows elaboration in memory in a way that enables better retrieval when the information is needed (Baddeley, 1990). Also, crews that critically analyze their own performance may 'buy into' the conclusions of their own analysis more deeply than they would accept the judgments of an instructor – especially since some crewmembers have as much flight experience as their instructors. Another potential benefit of facilitating self-debriefing by crews is that it promotes analytical skills and encourages crews to use those skills to analyze and discuss their own performance in line operations (Wolfe, this volume).

Although the concept of facilitated debriefings has been long established and endorsed formally by the Federal Aviation Administration (FAA), airlines, and other organizations, until recently its practice had not been studied systematically. Many questions were not addressed. Chief among those questions was: How effective is the facilitated approach to debriefing LOFT? The concept and techniques of facilitation were developed in settings very different from aviation, mainly adult education and business settings such as retreats in which managers discuss their organizational goals and issues (e.g., Gibb, 1982; Mills & Roberts, 1981). It is not clear how much the original concepts of facilitation must be modified to work effectively in aviation settings.

In private discussions, managers in several airline training departments expressed concern about how to train instructors to facilitate debriefings effectively. What little written material was available had been taken from materials for settings other than aviation with little adaptation. In most cases the material was couched in quite general terms and provided few detailed examples of how to use facilitation techniques in LOFT debriefings. Few airlines gave instructors hands-on practice and feedback; this shortcoming was especially problematic because facilitation is a skill quite different from the traditional techniques of aviation instruction.

Funded with a grant from the FAA, we began a field study in 1994 to address these concerns. Our study addressed five major questions:

1. To what extent do instructors attempt to facilitate crew participation and self-analysis in LOFT debriefings?
2. What techniques do instructors use to facilitate, and how effective are these techniques?
3. How much variation occurs among instructors and among airlines in the conduct of debriefings?
4. What is the character of crew participation, especially in terms of analyzing and evaluating their own performance?
5. Is facilitation a viable approach to encouraging crew participation and self-analysis?

*Techniques for Facilitation*

We reviewed several literatures relevant to facilitation: adult education, counseling, guidelines for facilitating business meetings, and CRM. Facilitation techniques relevant to LOFT debriefing can be grouped into four categories:

*Introductions* An explicit introduction is necessary to clarify the role of the facilitator and the nature of the participation expected of the group (Casey, Roberts, & Salaman, 1992; Nelson-Jones, 1992; Gibb, 1982). A good introduction can also motivate the group to participate by providing a rationale for the session.

*Active listening* Good listening skills enable the facilitator to work with what the participants are saying and to encourage further participation. Active listening shows that the facilitator is attending to the speaker, understands what is being said, and wants to hear more. Active listening can range from a simple 'uh-huh' or 'okay' to echoing or reflecting in one's own words what a speaker is trying to communicate.

*Questions* According to the Socratic method, learning is facilitated by questioning, encouraging exploration, and pushing for explanation; not by lecturing and giving the answers (Casey et al., 1992). 'Can you give me a specific example?' 'How did you and the other person actually behave?' and 'What were your thoughts in the situation?' are examples of questions that can aid self-assessment (Nelson-Jones, 1992). Mills and Roberts (1981) assert that, ideally, questions should be brief; open (i.e., non-restrictive, don't imply opinion or judgment); and begin with who, where, and when for factual responses or what, how, and why for more in-depth and detailed answers.

The use of probing questions encourages active and in-depth participation. Probing questions that ask participants to explain and justify their responses have been reported to be particularly effective (Jacobsen, Eggen, & Kauchak, 1989). Mills and Roberts (1981) identified seven types of probes that encourage continued participation: non-verbal (e.g., a nod); short verbal ('Uh, huh?'); 'W' words (especially what, how, and why); statements such as 'Tell me more'; echoing of participant words; reflection of what the participant said with different words but the same meaning; and specialized reflections that imply more than was stated by the participant (also see Eitington, 1989.)

*Silence* Group participants do not always respond immediately to a leader's question. Most people find silence in a group setting uncomfortable, and leaders often allow no more than a one-second pause before rephrasing a question or answering it for the group. However, one second may not be long enough for participants to formulate a thoughtful response. Studies show that waiting three to four seconds substantially improves both the number and quality of responses (Rowe, 1986; Jacobsen et al., 1989). The longer pause elicits longer, more confident responses from the group, as well as more numerous voluntary observations, participant interactions, and participant questions. Furthermore, responses from slower participants increase, speculative responses and evidence-inference statements increase, and failures to respond decline (Ornstein, 1990; Rowe, 1974).

Although these kinds of techniques are important tools for facilitation, most crucial is the extent to which an instructor understands and accepts the underlying concept of facilitation: It is the instructor's job to help the crew make their own analysis. To do this the instructor must focus on identifying what assistance and encouragement each particular crew needs in order to engage in self analysis. As we discuss later in this chapter, crews vary substantially in how prepared they are to debrief themselves; thus instructors must adapt their approach to the particular crew. Also, instructors sometimes feel conflicted because they have observed problems and issues in crew performance during the LOFT that they want the crew to discuss. It is easy for instructors to slip into focusing entirely on guiding the crew to the instructor's perspective or even to fall into outright lecturing. We will suggest ways that instructors can combine facilitating crew self-analysis with the instructor's own observations.

*Field Study*

We observed, made audio recordings from, and later transcribed the debriefings of 36 U.S. airline crews who were completing their annual recurrent LOFT training between June 1994 and May 1995 (see Dismukes, Jobe, & McDonnell, 1997, and Dismukes, McDonnell, & Jobe, 2000, for details of research methods and data analysis). The crews were from five large, well-established international companies; four were passenger airlines and one was a cargo company. Twenty-five crews were two-person, and 11 were three-person crews. At each airline we interviewed training department managers and instructors for general comments about LOFT debriefing and instructor training.

We analyzed the debriefings with two types of measures. Descriptive measures included a large number of variables characterizing the nature of instructor and crew participation and the content of discussion. The other type of measure was the Debriefing Assessment Battery (DAB), which we developed to assess instructor and crew participation (Appendix). The eight instructor items and the six crew items of the DAB correspond to the objectives of debriefing (Federal Aviation Administration [FAA], 1990) and the concepts of facilitation and adult learning (Dismukes et al., 1997). Inter-rater reliability of the DAB is quite good and it can be used in real time to evaluate instructor facilitation and crew participation.

## Results and Discussion

Although our study was not concerned with evaluating the overall competence of the instructors, the great majority were clearly highly competent instructors, were conscientious, and displayed strong interpersonal skills. All seemed comfortable with and committed to the concepts of CRM; nevertheless most were not highly effective as facilitators. We interpret this discrepancy in terms of instructor training, discussed later.

At all five airlines most debriefings did not occur immediately after the LOFT. Instead, after a short break, the instructor and crew first returned to the simulator to conduct about two hours of practice maneuvers as rehearsal for the proficiency check that would follow the next day. (Typically, LOFT is conducted in conjunction with FAA-mandated recurrent training and checking of proficiency in technical aspects of flying). A few instructors, apparently on their own initiative, reversed the order when scheduling allowed so they could debrief the LOFT first. On several occasions crewmembers spontaneously mentioned that they had trouble remembering relevant aspects from the LOFT; the delay in debriefing may have contributed to this difficulty. Performing the practice maneuvers before the debriefing, in the same simulator, and under similar conditions seems likely to interfere with memory of the preceding LOFT. Our data do not directly address this issue, and it deserves further study. The typical delay in debriefing seems to be driven by the economic need to optimize use of simulators. We suggest that it would be worthwhile to explore ways to arrange for debriefing to follow immediately after LOFT.

Within each airline scores of individual instructors and crews varied over a wide range for most variables. (Presumably, the variation in crew scores was partially driven by variation among instructors). This large variability has major implications for standardization and training, discussed later. Substantial differences occurred among airlines in average values for some variable, however most of these differences were not statistically significant because of the large variability within airlines.[1]

The average duration of the debriefings was 31 minutes, with a range of 8 to 82 minutes. This figure includes the time spent watching videos of the LOFT for the four airlines that used videos (typically about one-third of the total session was spent watching video segments). Half-hour sessions allowed group discussion of only a few examples of crew performance, and often did not provide adequate time for in-depth analysis. Given all that occurs in a typical two-hour LOFT and the importance of deep analysis of what happened and how the crew managed the situations confronting them, it seems highly desirable to spend more than 31 minutes on debriefing. Although our data do not indicate what duration would be optimal, a thorough discussion was often accomplished in debriefings lasting about an hour. Obviously, airlines must consider cost-benefit tradeoffs in the amount of time scheduled for debriefing. Instructors do need to vary the length of the session according to the training needs of the crew, but the ten-fold range of duration observed in this study is clearly problematic.

Most debriefings followed the same general format. The instructor either gave a very short introduction or none at all and then led a discussion of segments of the LOFT in chronological order. Rarely did the instructor engage the crew in setting an agenda for discussion, although some instructors invited general comments on the LOFT before starting discussion of specific segments. Among the eight DAB instructor items *Explains the Debriefing Process* scored lowest, with an average slightly above 2 ('marginal') on the 7 point scale and a range of 1 to 5.

On average, instructors talked much more than any of the crewmembers; however, individual instructors differed substantially (for example, the percentage of talking by instructors with three-person crews ranged from 17 to 87). Instructors clearly led the content and form of discussion. Discussion typically revolved around the instructors' questions and comments. Eighty percent of the utterance patterns consisted of an instructor question or statement followed by a usually brief response from a single crewmember, and then another instructor utterance.

Most instructors asked a large number of questions, averaging close to one question per minute. Questions appeared to be the instructors' primary method of guiding the discussion. Most crewmembers responded readily to questions, however, crew responses and self-initiated statements tended to be brief, and back-and-forth discussion among crewmembers was quite limited. Thirty-three percent of crew utterances were responses to instructor (or, rarely, crew) questions; 30% were self-initiated substantive statements that added content to the discussion; 30% were non-substantive statements, most of which involved maintenance of discourse (e.g., 'I see what you mean'); and 7% were questions. More than two-thirds of crew questions involved discourse maintenance or were extraneous rather than addressing content, raising new issues, or bringing new information into the discussion.

On average, roughly half of the content of the debriefings was directed to discussing the crews' performance (CRM, technical, or mixed) in the LOFT. Instructors devoted substantially more discussion to positive aspects of crew performance than to negative aspects or ways to improve performance. Nearly two-thirds of the crew utterances concerning performance were neutral descriptions of what the crew did, with much less discussion of positive aspects, negative aspects, or ways to improve. The content of crew utterances was highly correlated with the content of instructor utterances. Instructors may have emphasized positive aspects of crew performance appropriately to give positive feedback, however more discussion of why some things went well and others did not go so well and how to improve performance would have been useful.

Our data suggest that the character of crew participation in many debriefings was not optimal for learning and retention. The average of crew overall score 3.5 on the seven-point DAB scale. The six crew items showed roughly similar averages and ranges. We designed the crew items of the DAB on the premise that experienced pilots would learn best by analyzing the situations that confronted them in the LOFT, evaluating their performance, and interacting directly with fellow crewmembers (FAA, 1990). Direct interaction among crewmembers may have several benefits: it puts greater responsibility on the crew, which may enhance the depth of what they learn: and it provides them with an opportunity to practice debriefing as they would conduct it on the line. Also, instructors sometimes note that pilots will accept an observation from a fellow crewmember more readily than from an instructor. To varying degrees, the instructors led the crews through analysis and evaluation in a step-by-step fashion, and the crews undoubtedly benefited, but their participation was limited in depth and more reactive than proactive.

Some instructors used facilitation effectively; however, most showed substantial room for improvement. The average of the DAB instructor *Overall Score* was 3.4 (range 1.0 – 5.6). The most striking deficiency was that most instructors failed to explain to the crew at the beginning of the debriefing how the debriefing would be conducted, what role the instructor would play, and what the crew was expected to do. Other common mistakes included failing to pause (use silence) when the crew did not respond immediately to questions, keeping the discussion centered on the instructor instead of encouraging the crew to interact with each other, making long soliloquies, evaluating crew performance before eliciting crew self-evaluation, failing to push beyond superficial description of events, and not getting the crew to analyze why things went well.

Scores on the six crew items on the DAB were moderately to highly correlated with scores on related instructor items. The two crew items best predicted by instructor DAB scores were *Analyze in Depth* and *Focus on CRM. Analyze in Depth* was most strongly correlated with the instructor items *Encourages Crew Analysis* (.72) and *Overall Score* (.72). *Focus on CRM* was most strongly correlated with the instructor items *Focus on CRM* (.72) and *Overall Score* (.73). The two crew items least predicted by instructor DAB scores were *Proactive Participation* and *Interact with*

*Each Other*. *Proactive Participation* was most strongly correlated with the instructor items *Focus on Crew* (.50) and *Overall Score* (.48). *Interact with Each Other* was most strongly correlated with Encourages Crew Evaluation (.47). Crew Overall Score was strongly correlated with instructor *Overall Score* (.68) and *Use of Questions* (.68). Thus, roughly half of the variance in crew participation can be accounted for in terms of instructor use of facilitation techniques. These data support the argument that facilitation is an effective way to encourage and guide crew participation in debriefing.

Correlational data, of course, do not necessarily imply causality or direction of causality. It is conceivable that instructors sometimes encountered non-responsive crews and subsequently abandoned efforts to facilitate. However, our subjective observation was that this did not occur to any appreciable extent. Furthermore, aspects of the data argue against this interpretation. The DAB item *Explains Debriefing Process* addresses an aspect of facilitation that occurs before the crew begins participation; it was positively correlated with crew scores on *Analyze in Depth* (.40), *Focus on CRM* (.49), *Respond in Depth* (.49), and *Overall Crew Score* (.37). Furthermore, *Explains Debriefing Process* and all of the other items were negatively correlated with the amount of talking done by the instructors.

We interpret our overall data as indicating that instructors attempted to elicit crew participation but retained moment-to-moment control of the discussion, rather than attempting to assist the crew in conducting their own debriefing. Some instructors may have felt that it was appropriate or necessary to keep close control of the discussion. Facilitating crew discussion takes longer than lecturing to the crew, and some of the instructors may have been responding to time constraints. Other instructors might not have realized that they were dominating the discussion or might not have known how to get the crew to take a more proactive role. Clearly, these debriefings were not led by the crews, however our data do not reveal how effectively the crews might have led their own debriefings had they been charged to do so by the instructors.

## Implications for Training

The uniformly low instructor scores on *Explains Debriefing Process* indicate that this is an area in which instructors have not been adequately trained. It seems a matter of common sense to tell crews explicitly what is expected of them. What the instructor says in the introduction and how he or she says it sets the tone for the entire debriefing. A good introduction is easy to provide; thus, training departments may be able to improve crew participation with relatively little effort by emphasizing this topic to instructors. Ideally, the introduction should describe how the debriefing will be conducted, explain how the crew is expected to participate and what the instructor role will be, and provide an explicit rationale for the benefits of crew-centered debriefings.

The wide variability of instructor effectiveness in facilitation and emphasis on crew participation suggests that the airlines face an issue of standardization and quality control of debriefings. This variability may reflect inadequate training of instructors in the techniques of facilitation; however, we have no data on the amount or content of training individual instructors received. When interviewed, several instructors spontaneously volunteered that they did not feel adequately trained to facilitate. Facilitation, especially because it departs radically from the instructional techniques traditionally used in aviation, requires hands-on training in which instructors observe expert facilitators, practice facilitating, and receive feedback.

Most training materials for teaching instructors to facilitate debriefings are couched in very general terms and were adapted from facilitation in very different business settings. To provide more detailed guidance specific to LOFT debriefing, we published a facilitation training manual (McDonnell, Jobe, & Dismukes, 1997), which is excerpted in chapter 3 of this volume. Klair (chapter 6, this volume) has developed a specific technique for debriefing line incidents and has also used this technique for debriefing LOFT (Klair, 1997).

Instructors sometimes mistakenly assume that using facilitation requires relinquishing their role as teacher in the debriefing. This misunderstanding may arise because in some business settings, facilitators lack the deep subject matter expertise of the group members; in these settings the facilitator is primarily concerned with maintaining a process that will enable the group to work together effectively. In contrast, the LOFT instructor has deep subject matter expertise in both technical and CRM matters. Furthermore, from his or her position in the simulator cab the instructor can see things happen in the LOFT that the crewmembers are unable to observe. This expertise and these observations play an important role in the debriefing. Instructors who are effective facilitators learn how to integrate their observations into group discussions in which the crewmembers are full participants.

The importance of crews participating substantially in LOFT debriefings is widely recognized. Some authors have gone beyond this principle to argue that the crews themselves should lead the debriefings, using the instructor as a resource (see, for example, Butler, 1993). Only one of the instructors in our study attempted to have the crew lead the debriefing. Although that debriefing was one of the better ones in terms of the level of crew participation, the crew only partially understood what constituted a good debriefing and needed considerable help. Before crews can debrief themselves, they must be taught how to conduct a debriefing. This could be the subject of classroom training that precedes the LOFT. This classroom training might include videos of crews debriefing themselves effectively and should provide crews opportunities to practice debriefing.

With the one exception noted above, the debriefings we observed revolved around the instructor's questions and comments, even when the instructor used facilitation techniques effectively to elicit crew participation. Is it better for the instructor or the crew to lead the discussion? No data exist to answer this question. Also we are not aware of any model that explicitly describes how a crew should conduct a debriefing. Leading the discussion provides the instructor with greater moment-to-moment control of content, and still permits using facilitation to elicit substantial levels of crew participation. On the other hand, crews may learn more from questions they pose themselves; also, taking more responsibility for the LOFT debriefing may encourage crews to debrief themselves on the line. This issue deserves further exploration.

Realistically, many crews may not yet have the skills or motivation to lead their own debriefings without substantial assistance from the instructor. This situation might change over time, however, if instructors consistently encourage crews to take a proactive role in debriefing their own LOFT and to consider the benefits of debriefing after line operations.

At the current state of industry practice, instructors who attempt to encourage crews to self-debrief, or to at least take greater responsibility for the direction of the debriefing, will encounter widely varying levels of crew responsiveness. McDonnell, et al. (1997, and chapter 3, this volume), drawing upon a concept expressed by Continental Airlines (1992), suggest that facilitation can be conducted at levels ranging from low to high, depending on the level of initiative and the self-debriefing skill of the particular crew. In the highest level of facilitation the instructor assists the crew in their own analysis, blending his or her own observations into the group discussion. In low-level facilitation the instructor leads the debriefing, directs the crews attention to critical issues, and may need to lecture to insure points are understood, but the instructor still attempts to foster as much self-analysis as possible.

## Summary

Most instructors in this study attempted to facilitate crew participation, but their success ranged from very good to poor. Within each of the five airlines instructors varied widely in their effectiveness in facilitation and emphasis on crew participation. This suggests a need for better standardization within airlines.

Instructors used questions as their primary tool to elicit crew participation and guide the content of discussion. Instructors used other facilitation techniques less frequently but significantly influenced crew participation when they did. The content of crew discussion was strongly driven by the content on which the instructors focused. Crews did not lead their own debriefings, but did participate actively and responded positively. The concept of having crews debrief themselves using the instructor as a resource may not always be feasible at present, but it deserves further exploration and development.

The 31-minute average for debriefings in this study limited the depth and range of analysis of the LOFT. Crews would very probably learn more from the LOFT with a longer debriefing; however this issue involves cost-benefit tradeoffs that can only be evaluated by the airlines themselves.

It has been a matter of faith among training departments that facilitation is an effective tool to encourage crews to analyze their performance in LOFT in ways that will benefit them in line operations. This study provides empirical evidence that skilled facilitation works. Instructors who facilitated effectively elicited more participation and deeper analysis by crewmembers than did instructors who did not use facilitation effectively. Airlines would benefit by giving instructors explicit, hands-on training in facilitation, followed up with mentoring. Since the initial publication of the results of this study, several airlines have expanded their training for instructors in facilitation, and it seems likely that debriefing will improve.

## Acknowledgements

This work was funded by the FAA's Office of the Chief Scientist and Technical Advisor for Human Factors (AAR-100), Eleana Edens, program manager, and by the AOS Training Element, NASA Ames Research Center.

## Note

1 All differences discussed later in this paper were statistically different at the .05 level or better unless otherwise stated.

## References

Baddeley, A. (1990). *Human memory: Theory and practice*. Massachusetts: Simon & Schuster, Inc.

Butler, R. E. (1993). LOFT: Full-mission simulation as crew resource management training. In E. L. Wiener, R. L. Helmreich, & B. G. Kanki (Eds.), *Cockpit resource management* (pp. 231-259). San Diego: Academic Press.

Casey, D., Roberts, P., & Salaman, G. (1992). Facilitating learning in groups. *Leadership and Organization Development Journal, 13*(4), 8-13.

Continental Airlines, Flight Operations Human Factors Group. (1992, January). *LOFT facilitation techniques: Practical strategies and techniques for LOFT facilitators*. Unpublished manuscript.

Dismukes, R. K., Jobe, K. K., & McDonnell, L. K. (1997). *LOFT Debriefings: An analysis of instructor techniques and crew participation.* (NASA Technical Memorandum 110442). Moffett Field, CA: NASA Ames Research Center.

Dismukes, R. K., McDonnell, L. K., & Jobe, K. K. (2000). Facilitating LOFT debriefings: Instructor techniques and crew participation. *International Journal of Aviation Psychology, 10* (1), 35-57.

Duval, S., & Wicklund, R. A. (1972). *A theory of objective self awareness*. New York: Academic Press.

Eitington, J. E. (1989). *The winning trainer*. Houston: Gulf Publishing.

Federal Aviation Administration (1990). *Line operational simulations* (Advisory Circular 120-35B). Washington DC: Author.

Gibb, P. (1982, July). The facilitative trainer. *Training and Development Journal*, 14-19.

Helmreich, R. L., & Foushee, H. C. (1993). Why crew resource management? Empirical and theoretical bases of human factors training in aviation. In E. L. Wiener, R. L. Helmreich, & B. G. Kanki (Eds.), *Cockpit resource management* (pp. 3-41). San Diego: Academic Press.

Jacobsen, D., Eggen, P., & Kauchak, D. (1989). *Methods for teaching: A skills approach* (3rd ed.). Columbus: Merrill Publishing Company.

Klair, M. (1997). Briefing and debriefing techniques for instructors in aviation. In R. S. Jensen & L. A. Rakovan (Eds.), *Proceedings of the Ninth International Symposium on Aviation Psychology* (pp. 568-573). Columbus, OH: Ohio State University.

Lauber, J. K., & Foushee, H. C. (1981). *Guidelines for line-oriented flight training*. (NASA Conference Publication 2184). Moffett Field, CA: NASA-Ames Research Center.

McDonnell, L. K., Jobe, K. K., & Dismukes, R. K. (1997). *Facilitating LOS debriefings: A training manual.* (NASA Technical Memorandum 112192). Moffett Field, CA: NASA-Ames Research Center.

Mills, P., & Roberts, B. (1981, March). Learn to guide and control discussion. *Successful Meetings*, 96-97.

Nelson-Jones, R. (1992). *Group leadership: A training approach.* Pacific Grove, CA: Brooks/Cole Publishing Co.

Ornstein, A. C. (1990). *Strategies for effective teaching.* New York: Harper & Row.

Rowe, M. B. (1974, February). Wait-time and reward as instructional variables. *Journal of Research in Science Teaching*, 81-97.

Rowe, M. (1986, January/February). Wait-time: Slowing down may be a way of speeding up. *Journal of Teacher Education*, 43-50.

Slamecka, N. J. & Graf, P. (1978). The generation effect: Delineation of a phenomenon. *Journal of Experimental Psychology: Human Learning and Memory*, *4*(6), 592-604.

Smith, G. M. (1994). *Evaluating self-analysis as a strategy for learning crew resource management (CRM) in undergraduate flight training.* Ann Arbor, MI: Dissertation Abstracts.

# 3 Facilitating LOS Debriefings: Synopsis of a Training Manual

LORI K. MCDONNELL, KIMBERLY K. JOBE, AND R. KEY DISMUKES

This chapter is excerpted from our facilitation training manual (McDonnell, Jobe, & Dismukes, 1997), which was developed as a practical guide to help airline instructors facilitate debriefings of Line Operational Simulations (LOS) effectively.[1] It is based on our study of Line Oriented Flight Training (LOFT) debriefings at several U.S. airlines (Dismukes, Jobe, & McDonnell, 1997), which is summarized in chapter 2 of this volume. This chapter will present specific facilitation techniques instructors can use to achieve debriefing objectives, which are derived from the literature cited in the first two chapters and our research findings presented in chapter 2. This guidance is designed to be flexible so it can be tailored to the individual needs of each airline. Part 1 clarifies the purpose and objectives of facilitation in the LOS setting. Part 2 provides recommendations for clarifying roles and expectations and presents a model for organizing discussion. Part 3 suggests techniques for eliciting active crew participation and in-depth analysis and evaluation. Finally, in Part 4, these techniques are organized according to the facilitation model.

## Part 1. An Introduction to Facilitation

The facilitation techniques presented in this chapter are intended to provide instructors with a set of tools they can use to facilitate crew-centered debriefings effectively. These tools supplement, rather than replace, the skills instructors already possess. Becoming skillful at facilitation requires practice, but once mastered it enables the instructor to increase what crewmembers learn and take back to the line.

Effective facilitation enables the instructor to meet the following objectives of crew-centered debriefings:

- Crewmembers discuss issues directly with each other (rather than interacting solely with the instructor) and discover as much on their own as possible.

- The crew thoroughly analyzes and evaluates what happened in the LOS, how they handled the situation, what went well, what could be improved, and how to improve it.
- The crew recognizes how CRM techniques helped them manage or could have helped them manage the situations they encountered in the LOS.
- Crewmembers leave the session with a better understanding of how they can use CRM on the line to enhance safety and efficiency.
- The crew is encouraged to develop the habit of self-debriefing following line operations.

**Instruction vs. Facilitation**

In a crew-centered debriefing, the objective is for the instructor to facilitate crew discussion so the crewmembers do most of the talking, participate proactively, discuss issues with each other, and thoroughly analyze the situations that confronted them and how they managed those situations. Some of the CRM literature espouses the ideal that crews debrief themselves and use the instructor as a resource. Realistically, most crews lack sufficient experience in analyzing the CRM aspects of cockpit operations to be able to conduct their own debriefings without assistance. Therefore, instructors need to facilitate each crew's analysis of their LOS performance and avoid centering the debriefing on their own perceptions. Crews vary considerably in how well they analyze what happened in the LOS and how well they respond to encouragement to participate actively; thus, instructors must adapt the level of facilitation to the capabilities of each crew. (The levels of facilitation are discussed in detail later in this chapter.)

The concept of facilitation was originally derived from settings outside of aviation. In some of those settings the facilitator would act only as a moderator of discussion and lacked the technical expertise held by the group members. In the LOS setting, however, the instructor plays the dual role of facilitator *and* instructor. Gibb (1982) calls this the *facilitative trainer*. By using facilitation, instructors enable crews to figure things out for themselves while using their own expertise as an instructor to enhance the crew's understanding of the points they missed on their own.

Facilitation does not require that instructors completely withhold their own perspectives. From 'the back of the cab' instructors often see things the crew does not notice and can share their experience in how CRM can be used to manage the specific events that occurred in the LOS scenario.

Since crews learn better when guided to self-discovery, the instructor should use facilitation to promote that self-discovery in the crew and to encourage the crewmembers to analyze their performance to the fullest extent possible. Once crewmembers have completed their analysis, the instructor can reinforce the things the crew did well. For the points crewmembers miss, the instructor can provide the instruction necessary to ensure the training objectives are met.

Additionally, at times the instructor may need to correct erroneous statements made by a crew and/or provide specific information the crewmembers lack. This is better done by direct instruction than through facilitation, because facilitation only works when the participants are knowledgeable of the information being discussed. The instructor can either integrate brief instructional comments into the crew's discussion or provide instruction as needed after crewmembers have finished their analysis.

When guiding a crew's analysis, the instructor should avoid creating the impression of leading the crew to predetermined answers. The crew will not be motivated to participate unless the instructor conveys that the topics, issues, and perceptions the crewmembers raise are important.

*How to Facilitate the Debriefing*

- Set expectations for crew participation.
- Guide the session to the extent necessary to achieve the debriefing objectives.
- Adjust facilitation to the level needed to engage the crew to the maximum extent possible.
- Draw out quiet crewmembers.
- Ensure that all critical topics are covered.
- Integrate instructional points as needed into the crew's discussion.
- Reinforce positive aspects of the crew's behavior.

*What to Avoid*

- *Avoid lecturing and having the debriefing become an instructor-centered session.* Crew participation is hindered when instructors give long lectures and drive the session with their own opinions. Avoiding lecturing does not mean instructors cannot teach, but they should teach through facilitation by integrating their own perspectives into the crew's discussion. Rather than telling the crew what they did wrong during the LOS and how they can improve, the instructor should try to get the crew to figure this out for themselves. When crews discover what they need to work on by themselves, they are much more likely to learn from their mistakes and carry that learning over to the line.

- *Avoid providing analysis and evaluation of crew performance before the crew have completed their own analysis.* It is important that crewmembers learn to analyze and evaluate their own performance so they can develop the habit of debriefing their performance following line operations. If the instructor analyzes and evaluates crew performance before the crewmembers have a chance to do so themselves, the crew will feel less responsible for making their own analysis. The crew can benefit more if they are encouraged to analyze and evaluate their own performance first. Then the instructor can supplement and reinforce crew analysis and evaluation as appropriate.
- *Avoid giving the impression that only the instructor's perceptions are important.* The instructor needs to make it clear that what the crew has to say is important so the crew will be encouraged to participate actively and analyze and evaluate their performance in depth.
- *Avoid interrupting crew discussion.* Interrupting shows a lack of interest in the crew's views and may disrupt their train of thought and hinder their process of self-discovery. Interruptions also undermine the crew's sense of responsibility for making their own analysis and evaluation. Rather than interrupting crewmembers, the instructor should make a note of any concerns and bring up these issues after the crewmembers have completed their comments. However, interrupting may be necessary if a crewmember makes a statement contrary to Federal Aviation Regulations or company Standard Operating Procedures (SOPs).
- *Avoid interrogation.* In order for the crew to be able to take responsibility for initiating topics and discussing issues interactively, the instructor must work to keep the discussion crew-centered. Asking questions in a manner that resembles interrogation rather than guidance can hinder crew discussion, forcing the crew into a reactive, rather than proactive, role.
- *Avoid having a rigid agenda.* The instructor should enlist the crew's help in developing the agenda in order to encourage them to identify areas in the LOS where they did well and/or need improvement. Instructors should keep the focus of the debriefing on topics introduced by the crew, and introduce their own points as necessary to expand on crew-initiated topics and to cover important issues not initiated by the crew.
- *Avoid shortchanging high-performing crews by cutting their sessions short.* The instructor should not assume that because a crew did well there is little to talk about, so a shorter session will suffice. Crews that

perform well may not know why the LOS went well. Good crews can learn from analyzing what they did well and *why* it worked. Instructors need to make sure that these crews recognize and discuss which CRM techniques resulted in their positive outcomes so the crewmembers can apply this knowledge in future situations, when flying with different crewmembers on the line.

## Levels of Facilitation

The instructor's role in the debriefing session will vary as a function of the level of facilitation needed for each crew. Ideally, the crew will be able to analyze and evaluate their own performance and the instructor will need to give only minimal guidance, thus utilizing high-level facilitation. Unfortunately, this rarely happens in the current training environment because crews lack the experience to participate at this level. As a result, the instructor must adapt to the capabilities of each crew. Currently, most debriefing sessions are conducted at an intermediate level of facilitation. These sessions are largely driven by the instructor, who leads the crew to important topics and utilizes questions to evoke in-depth analysis and evaluation by the crew.

As suggested by Continental Airlines (1992), the level at which facilitation is conducted is a broad continuum from high (most desirable) to low (least desirable). The instructor may need to adjust the level of facilitation to whatever level is appropriate throughout the session and should always strive to work at the highest level of facilitation at which each crew is able to respond. If facilitation takes place at too low of a level, the crew will be prevented from participating to their full capability. A sign of skillful facilitation is being able to recognize what level is appropriate for each crew, as well as being able to adapt to the varying needs of individual crewmembers as the debriefing discussion shifts from one topic to another.

If instructors consistently use facilitation to generate crew-centered debriefings, crews will gain experience as they go through recurrent LOS and they will participate at increasingly higher levels each year.

### *High-Level Facilitation*

High-level facilitation is possible when the crew discovers and discusses important issues on their own with little guidance. Debriefing at this level provides the deepest and most enduring learning because the crewmembers guide their own self-discovery. For facilitating at this level, the crewmembers must be able to identify important topics and issues that arose during the LOS, set an agenda for discussing these issues, and critically analyze the situation they faced and evaluate how well they performed.

At this level of facilitation, The instructor's role is to inform the crew of the objectives for the debriefing, outline the debriefing process, and assist in guiding the discussion only when necessary. The instructor, who has an advantaged position as an experienced observer, can serve as a resource for the crew to help reinforce crew observations and supplement their perceptions. Although the instructor retains responsibility for ensuring that the debriefing objectives are met, this should be achieved through general guidance rather than leading the discussion moment-to-moment.

*Intermediate-Level Facilitation*

When crewmembers are not as adept at conducting their own analysis and evaluation of the LOS, the instructor must lead the discussion more directly. At the intermediate level of facilitation the instructor may have to help the crew discover important issues and lessons by asking questions which lead them to specific issues and topics. The instructor may also need to encourage the crew to analyze the situation and their performance in greater detail. At this level, crewmembers are able to achieve self-discovery through analysis and evaluation of their performance with moderate guidance. The instructor may need to supplement crew analysis substantially, but should do so only after the crewmembers have completed their own analysis.

*Low-Level Facilitation*

Low-level facilitation does not imply inadequate facilitation on the instructor's part; rather, it is the necessary and appropriate level to use when crews do not respond to higher levels. When crewmembers show little initiative or respond only superficially to the instructor's efforts to get them to analyze the LOS situations and evaluate their own performance, the instructor may be forced to be more direct and to lead the discussion step by step. However, few crews are truly incapable of participating at a higher level, and before resorting to this level of facilitation, instructors need to make sure they are not unwittingly discouraging crew participation (as described earlier under practices to avoid). At this level, self-discovery by the crew is limited but the instructor should still use facilitation techniques to lead the crew to critical issues, appropriate solutions, and correct evaluation. At the end of the discussion of each topic the instructor may have to explicitly summarize the nature of the problem encountered in the LOS and describe how it should be handled. When low-level facilitation is necessary, the instructor should continue trying to elicit crew participation at a higher level. If it becomes necessary to instruct the crew on important points, it is important to confirm that the crewmembers understand and agree, and to continue to encourage them to initiate further discussion on their own.

## Criteria for Effective Crew Participation

In order to determine the appropriate level of facilitation to use, the instructor needs to be aware of how capable the crewmembers are of participating. The following criteria can be used to determine what level is appropriate for each crew at various times during the debriefing. If the crewmembers meet all of these criteria, high-level facilitation is best. If not, intermediate- or low-level facilitation may need to be used.

- Crewmembers *analyze* in depth their LOS performance, discussing the situations they confronted, what they did to manage the situations, and why they made the decisions and performed the actions they did.
- Crewmembers *evaluate* in depth their LOS performance, discussing what went well, what did not go well, and how their performance could have been improved.
- Crewmembers discuss how CRM techniques helped them manage or could have helped them manage the situations they encountered in the LOS.
- Crewmembers address each other directly and interactively discuss the LOS rather than merely respond to the instructor's direction.

## Criteria for Effective Instructor Facilitation

The instructor's effectiveness in facilitating the debriefing can be measured according to how consistently and efficiently the following criteria are met:

- Facilitates crew discussion at the highest level possible, guiding the discussion only to the extent necessary to help the crew accomplish the objectives of the debriefing.
- Modifies the level of facilitation to accommodate the crew's needs throughout the debriefing. Refrains from interfering with crew discussion if the crew does not need guidance. If the crew needs

guidance, uses intermediate-level strategies to elicit continued and/or deeper discussion and reverts to low-level facilitation only if necessary to ensure important lessons are learned.

- Ensures that the crewmembers analyze their performance in depth, identify CRM techniques that played or should have played a role in the LOS, and discuss the LOS with each other in an interactive way.
- Ensures that all critical topics are covered.

## Part 2. Getting Started

### Clarifying Roles and Expectations: The Introduction

The purpose of the introduction is to provide a persuasive rationale for crew-centered debriefing. The instructor needs to explicitly state how the crew should participate in the debriefing and how the instructor will participate.

A good introduction is very important to encourage the crew to take responsibility for the debriefing session. The crew can't be expected to participate in a certain way if they are not told *how* they are expected to participate – individuals want and need to know what is expected of them. Because introductions clarify expectations, they can help increase the amount and depth of crew participation.

Effective introductions should address all of the issues in the following four categories:

*The Instructor's Role*

- Outline the process for the debriefing and help establish an agenda.
- Help the crew identify topics and facilitate crew discussion as necessary.
- Act as a resource for CRM and technical issues.
- Keep discussion crew-centered rather than instructor-centered.
- Ensure that the training objectives are met.

*The Crew's Responsibilities*

- Raise issues and initiate discussion.
- Discuss issues directly with each other rather than only with the instructor.
- Discuss CRM techniques used during the LOS and other techniques

which could have been used profitably.

- Critically analyze the situations that were encountered and how they attempted to manage those situations.
- Evaluate how things turned out and why.
- Discuss what, if anything, they would do differently in the future.

*The Rationale for Using Crew-Centered Debriefing*

- Crews learn better through self-discovery and self-analysis than by lecture.
- It draws upon crews' professional expertise and motivation to perform well.
- It helps the instructor understand the crew's performance.

*The Expected Length and Format of the Debriefing*

- Tell the crew how long the session will last (so they will not be motivated to keep quiet in hopes of a short session). A thorough crew-centered debriefing can usually be accomplished in an hour.
- Tell crews explicitly what format will be used for the debriefing (e.g., setting an agenda of topics to discuss, using videos to introduce and supplement discussion, using the model presented in this chapter to keep focused on important topics).

## Debriefing Format

*Developing an Agenda for the Discussion*

A primary debriefing objective is for discussion to be crew-centered. The instructor plays a critical role in helping the crew develop a structure for their discussion and in ensuring that all important issues that arose during the LOS are covered. It is important to get the crewmembers to actively participate in developing the agenda so they can develop the skill of identifying issues that arise in line operations.

The first step is to ask the crew to identify any topics, situations, or performance issues they would like to discuss. The crew should be encouraged to include aspects of their performance that worked well, not just aspects that could be improved. A good way to improve the ability of crews to identify issues for discussion is to suggest during the pre-LOS briefing that they make note of issues that arise in the LOS that they would like to discuss in the debriefing.

The issues the instructor noted during the LOS are also an important part of the debriefing agenda. The instructor needs to ensure that all critical issues are covered and can introduce issues not identified by the crew at an appropriate point during the debriefing.

Typically, it is easiest to discuss phases of the flight in chronological order so videotaped segments of the LOS performance can be incorporated into the discussion without requiring extensive searching on the VCR. However, this is largely a matter of personal preference and some instructors prefer to start the discussion immediately when crewmembers suggest a topic in order to reinforce the crew's initiative. One advantage of showing video segments related to each topic is that after viewing their performance objectively, crewmembers may be better able to analyze and evaluate their performance in depth.

### *Organizing the Discussion: The C-A-L Model*

The C-A-L Model incorporates three main concepts: CRM (C), Analysis and evaluation of LOS performance (A), and application to Line operations (L). This model provides a simple framework for crew discussion of each LOS situation or topic. Posting a copy of the model (Figure 3.1) where the crew can see and refer to it will help remind them of each aspect of their performance that they should address. Part 4 of this chapter demonstrates how to put this model to work.

## Part 3. Facilitation Techniques

Part of the instructor's role is to ensure that the crewmembers analyze and evaluate LOS situations and their performance in depth. One of the greatest difficulties the instructor may encounter is getting the crew to go beyond a superficial description of LOS events. The objective is to encourage crewmembers to not just discuss what happened – they should discuss *why* they did what they did to manage the situation and *why* the situation turned out the way it did. Discussing *why* helps crewmembers better understand what knowledge and information played a role in their decision-making process, which in turn can help them learn to make better decisions in the future.

This section will demonstrate how to effectively use the techniques that are essential to successful facilitation. The next section (Part 4) will show how to integrate these techniques utilizing the C-A-L model.

| | |
|---|---|
| **C** | **CRM - Applying the Company Model**<br>Use wallboards with list of CRM concepts.<br>✈ Tie CRM concepts and techniques to operational issues.<br>Put CRM into practice.<br>✈ Crew discussion of the LOS should be interactive. |
| **A** | **Analysis and Evaluation of LOS Performance**<br>Explicitly evaluate performance during the LOS.<br>✈ How effective was management of the situation?<br>- What went well, and why?<br>- What could be improved, and how?<br>Interactively analyze the situation confronted.<br>✈ What happened?<br>✈ How was it managed (include CRM techniques utilized)?<br>✈ Why was it managed that way? |
| **L** | **Line Operations - Applying Lessons from LOS**<br>Discuss how the LOS performance and associated CRM issues relate to line operations.<br>✈ Discuss related line incidents that illustrate CRM issues.<br>✈ Discuss how to apply LOS success to line operations.<br>Discuss how things could have been done differently.<br>✈ What could have been done differently to improve the outcome in the LOS?<br>- What CRM techniques could have helped?<br>- How could you turn *areas for improvement* into *strengths?*<br>✈ What can be done to prevent or manage similar situations on the line? |

**Figure 3.1 C-A-L Model for LOS Debriefings**

## Questions

Questions are useful to guide the crew to important topics and encourage them to analyze and evaluate their LOS performance in-depth. Although questions are one of the instructor's most useful tools, instructors should be wary of dominating the discussion with their own questions. To the extent necessary, the instructor can use questions to help the crewmembers identify issues and follow up on topics, and to encourage them to thoroughly analyze their performance through direct discussion with each other. Ideally, questions should be phrased to require more than a simple 'yes' or 'no' answer; questions should push the crew to analyze their performance at a deeper level.

### *Set the Scene and Ask for Crew Reaction*

A good way to start a discussion of a LOS event is to show the video clip of the event and/or describe the situation, and then ask an open-ended question: 'What went well (or not so well) there?' or 'What lessons can be learned from this?'

If the crew does not respond, try again with a more focused question: 'Is there anything you would do differently if you had it to do over?' or 'Did you see anything in the video that was not SOP?'

### *Lead the Crew to Topics*

If the crewmembers seem unsure of what to discuss next, or if they overlook an important aspect of a situation, the instructor may need to ask a question to focus their attention on the particular topic or aspect of their performance. It is important to phrase the question so it is not easy to answer with a simple 'yes' or 'no' and so it does not leave the crew with nothing to say.

### *Deepen the Discussion*

Once the crew has begun to discuss a topic, the instructor may need to encourage them to explore in more depth what happened. The instructor needs to ask questions that require the crewmembers to explore their thoughts and actions in the LOS, and avoid asking questions that give the answer or leave little for them to say.

*Follow Up on Crew Topics*

It is important to follow up on topics initiated by the crewmembers and to encourage the crew to explore their topics fully. Following up on crew topics demonstrates to the crew that their thoughts and ideas are important and reinforces their initiative. To obtain greater depth and detail on topics initiated by the crewmembers, the instructor needs to ask questions that begin with *what, how,* and *why.*

*Turn Crew Questions and Comments Back to Them*

If a crewmember makes a comment or asks a question, the instructor needs to avoid using this as a springboard for presenting his or her own perspectives. Instead, the instructor should continue to encourage the crew to work out answers for themselves. Questions are a powerful tool for blending the instructor's observations into the discussion without dominating the session. After the crewmembers have addressed everything they can, the instructor can add teaching points the crew has not yet discovered.

*Get All Crewmembers to Participate Actively*

It is important to ensure that all members of the crew actively participate in the debriefing so they can all experience the higher level of learning that results from taking an active role in the learning process. Also, the crew cannot fully analyze and understand their performance as a team without each member's perspective. If one or more crewmembers do not volunteer comments and respond only minimally to questions, the instructor can use several techniques to draw out quiet crewmembers and encourage active participation by the entire crew:

- *Ask the quiet crewmember to discuss how effectively the crew performed in the LOS.* It is relatively easy to respond to this type of question because there is no right or wrong answer and all crewmembers are likely to have an opinion.
- *Re-direct a question to the quiet crewmember.* If questions are predominantly answered by the same crewmember, repeat the question after the dominant crewmember has responded so that the quiet crewmember will be required to participate and communicate his or her opinions and ideas.
- *Ask the quiet crewmember to expand on what another crewmember said.* If a crewmember sits silently while another does all of the

talking, it may be helpful to ask the silent one to expand on what was said by discussing a new aspect of the topic. This technique will make it easier for the quiet member to find something to say.

- *Start with junior crewmembers.* To instigate crew participation, it may be helpful to start with the more junior members so they will not be put in the position of potentially disagreeing with the captain early in the session (see Klair, 1997, and chapter 6, this volume).

**Use of Silence**

Silence can be a very useful tool to elicit thoughtful crew responses. Most people are uncomfortable with silence in a group setting and are more likely to make some kind of response the longer the silence lasts. Without realizing it, instructors often allow only a second or two for a response to their question before either calling on someone else or giving the answer. However, one second is generally not long enough to formulate a thoughtful response, especially to complex questions.

*Benefits of Using Silence*

Studies show that waiting three to four seconds after asking a question greatly improves the number and quality of responses. The specific benefits of pausing at least three seconds include:

- Crewmembers produce longer and more confident responses.
- Crewmembers are more likely to volunteer unsolicited responses.
- Crewmembers interact more with each other.
- Crewmembers ask more questions.
- Less articulate crewmembers participate more.

*What to do During Silence*

In order to minimize the discomfort often associated with even brief periods of silence, it is important to behave in a way that makes the crew feel at ease.

- *Look relaxed, not anxious or impatient.* For the crew to feel comfortable with silence, it is important that the instructor appears to be at ease. Appearing tense and anxious will give the crew clear non-verbal cues that silence is not appreciated.
- *Sit back.* When instructors sit on the edge of their seat, crews may

interpret this as a sign of impatience. By sitting back in their seat, instructors demonstrate that they are comfortable allowing periods of silence for the crew to think before they respond.
- *Smile.* Smiling conveys reassurance and acceptance, which can help put the crew at ease.

*Strategies for Using Silence*

Silence can be used throughout the session to encourage crew participation, not only after asking a question but also after crew comments. Using silence is a high-level facilitation strategy – if silence alone does not elicit crew discussion, the instructor can try the next level of facilitation (see Troubleshooting at the end of Part 3).

- *Pause after asking questions to allow crewmembers to formulate thoughtful responses.* When crewmembers do not respond immediately, the instructor needs to be patient and remain silent rather than answering for them.
- *Pause after crew comments to allow crewmembers to resume discussion.* When there is a lull in crew discussion, the instructor again needs to be patient and wait for the crew to say more rather than immediately providing input.

## Active Listening

Good listening skills are as important to instructors as good speaking skills. Active listening is one of the most useful tools instructors can use to encourage continued participation; it shows the crew that they are listening and paying attention to them and that they understand what the crew is saying, which in turn encourages the crew to keep talking. Active listening ranges from simple non-verbal gestures to expanding on crew comments. In all cases, the objective is to encourage the crew to keep talking and, in the more complex forms, to encourage deeper discussion of the issues. The following are examples of the various levels of active listening, in order from the simplest to the most complex.

- *Non-verbal.* Nod, smile, make eye contact, lean forward in the chair, or otherwise indicate an interest in what the crew is saying.
- *Short interjections.* Interject with 'Yes?', 'Uh-huh…', 'I see…', etc. This indicates that the instructor is following what the crew is saying and encourages them to say more.

- *Echoing.* Repeat part of what the speaker said as a question directed back to the crew (e.g., 'So you weren't sure if you were cleared for takeoff?').
- *Reflecting.* Repeat what the speaker said in different words while retaining the same meaning (e.g., if speaker says 'We didn't use good communication' the instructor might respond with 'You didn't let each other know what you were doing?').
- *Expanding.* Expand on what the speaker said by implying more than the speaker intended (e.g., if the speaker says 'We didn't use good communication' the instructor might respond with 'So if you had communicated better you could have avoided getting overloaded?').

With all of these active listening techniques, the most important aspect is the tone the instructor conveys to the crew. It must be apparent from the instructor's tone that he or she is genuinely interested in what the crew has to say. Furthermore, the instructor must convey an interest in the crew's perspectives, not just in leading them to predetermined answers.

## Use of Video

The videotape of the LOS session is a useful tool for enhancing the debriefing. Relevant events need to be marked on the tape during the LOS session so the instructor can draw the crew's attention to important points for discussion.

### *Benefits of Using Video*

An important contribution of videos may be to remind crews what transpired at critical segments during the LOS because it is often difficult for participants to remember specific details of the typically long and stressful sessions. Videos enable crews to review their actions and then examine and gain insight into their behavior with the detachment of observers. Videos can also encourage self-assessment by helping crewmembers identify behaviors that were either helpful or harmful to the outcome. The realistic feedback that videos provide can make it easier for the instructor to involve the entire crew in a discussion of team performance, thus enhancing the overall potential of the training session.

### *Techniques for Using Video*

- *Index important events.* The instructor, who has experience with the LOS scenario, will be aware of critical events that will challenge the crew. The instructor should index these events on the videotape during the LOS and make note of important points to discuss. Noteworthy examples of good and poor crew performance should also be indexed.

- *Only show video segments intended for discussion.* The purpose of showing the video is to enhance crew analysis by refreshing their memories and giving them the opportunity to objectively evaluate their performance. If the instructor shows a video segment without discussing it, this wastes both time and the opportunity for discussion.
- *Avoid showing a large number of segments or very long segments.* In a one-hour session, 3–6 segments are usually all there will be time to adequately discuss. Instructor should select short segments that illustrate major aspects of crew performance (good or bad) rather than showing an entire phase of flight (e.g., from top of descent to touchdown). For aspects of a crew's performance that cannot be clearly illustrated by the video, it is best to steer the discussion to these topics without introducing them with a video segment.
- *Be proficient in the use of the video equipment.* Instructors should be very familiar with the video equipment so they can quickly find the events marked on the tape. Using the automatic index marker to find events is much more efficient than playing the video in fast forward while searching for segments.
- *Introduce each video segment.* For each segment the instructor needs to set the scene by describing the situation and where in the flight the segment to be discussed occurs. The instructor may also find it useful to remind the crew that they will be expected to analyze what they see.
- *Pause for comments.* The video should be paused whenever the instructor or a crewmember makes a comment, so the comment can be heard and understood. Crewmembers will feel that the instructor is genuinely interested in what they have to say if the instructor pauses the video and gives them undivided attention.
- *Pause to discuss important issues.* The instructor can pause the video to call attention to and elicit crew discussion of a specific aspect of crew performance shown on the video. For example, rather than talking during the video and forcing the crew to compete with the noise, stop the video and encourage crew discussion.

## Troubleshooting: When the Crew Does Not Participate Actively

Sometimes all or most of the crewmembers are unresponsive, and getting them to participate seems like pulling teeth. When this happens, instructors should first review their own participation to ensure that they are not unwittingly sending covert messages that discourage crew participation.

If crewmembers do not participate at the desired level, the following techniques may be helpful in eliciting more active participation. Start with the highest level of facilitation and progressively work down only through the levels necessary, so that facilitation stays at the highest level possible.

*High-Level Facilitation*

*Use silence/pauses to elicit crew responses.* Pause at least three to four seconds after asking a question rather than immediately answering for the crew. Most crewmembers will feel motivated to say something to end the uncomfortable silence. Pausing after crew comments could also encourage the crew to say more.

*Intermediate-Level Facilitation*

*Reword questions rather than give answers.* When crewmembers do not answer a question after five to ten seconds of silence, it may be necessary to rephrase the question to make the intent of the question clearer or to ask a more specific question. If they still do not answer, again pause to allow them time to formulate a thoughtful response. Rephrasing the question may give the crewmembers the cue they need to spark a response. By being persistent and patient, the instructor reinforces that the crewmembers are responsible for their own learning and that they are not going to get away with passively sitting by and waiting to be told the answers. When the instructor answers for the crew or rephrases questions in a way that leaves little for the crew to say, it inhibits in-depth crew participation and hinders crew learning.

*Ask crew to comment or elaborate on what was just said.* If the crew does not resume discussion after a period of silence, encourage further discussion by asking them to elaborate on their previous discussion.

*Low-Level Facilitation*

*Answer for the crew, but confirm that they understand and agree.* If the above facilitation techniques still do not elicit a response from the crew, or if the response given lacks depth, it may be necessary to briefly instruct the crew to ensure that they learn from the situation so they can apply the lesson on the line. If it becomes necessary to instruct the crew on certain issues, be sure to verify that they understand and agree because if they don't, the opportunity for learning may be lost. It is important to continue trying to involve the crew as much as possible. Once instructors have shared their knowledge with the crew and elicited their agreement, they need to try to return to a higher level of facilitation. This can be accomplished by asking for the crew's opinion, asking them to expand on the issue, or asking them to give an example of what was discussed from their line experience.

*Reinforce what the crew said.* If the crew does not resume discussion after a lengthy silence and cannot elaborate further, use the opportunity to provide the crew with positive feedback on the issue they were discussing. Not only will the feedback reinforce the issue at hand; it will also reinforce crew discussion in general.

## Part 4. The C-A-L Model in Action

The C-A-L Model provides a way to structure the debriefing. The first section, *CRM*, suggests strategies for helping the crew focus on CRM techniques that played a role in the LOS. The second section, *Analysis and Evaluation*, shows how to guide crews to identify and evaluate aspects of their performance that went well or could use improvement, including analysis of why the crew did what they did and why things turned out the way they did. The third section, *Line Operations*, provides a structure for helping the crew explore how they can apply what they learned from their analysis of the LOS to line operations.

A good way to organize debriefing each segment of the LOS is to show the appropriate video segment and then use the components of the C-A-L Model to guide the discussion.

### C: CRM – Applying the Company Model

The major purpose of LOS is to give crews a chance to practice using CRM concepts and techniques in realistic flight scenarios. Typically, crewmembers are much better prepared to talk about the purely technical aspects of a flight situation than the CRM aspects so they may need to be guided toward discussion of CRM issues pertinent to their performance in the LOS. Rather than discussing CRM as a set of abstractions, instructors can help crews the most by guiding them to consider how specific CRM techniques can be used to manage various flight situations.

*Focusing on CRM*

Instructors can use several techniques to bring CRM into the crew's discussion in a concrete, relevant way.

- *Refer to the posted CRM concepts.* Each company teaches a specific framework for CRM concepts that reflects the company's philosophy. Posting this conceptual framework on wallboards and referring to it during the discussion can help the crew relate CRM concepts and techniques to specific operational situations.
- *Use CRM-specific questions.* Questions that direct the crew to specific CRM topics can help the crew focus on and explore these CRM issues and how they presented themselves during the LOS.
- *Use guiding questions.* Guiding questions can be used to lead the crew to more specific and in-depth aspects of their LOS performance.

*Reinforcing the Utilization of CRM through Crew Interaction*

Instructors and crews too often fall into a pattern of discussion that centers entirely on the instructor: The instructor asks a question, a crewmember responds, the instructor comments, and the cycle repeats. The crew will benefit in several ways, however, if the instructor can get them to discuss their performance in the LOS directly with each other. Interactive discussion between crewmembers during the debriefing allows them to practice CRM skills such as communication and problem solving. It also leads them toward the ultimate goal of being able to debrief themselves in line operations. The instructor can counter the crewmembers' natural tendency to direct their comments to the instructor, rather than to each other, by using the following techniques:

- *Ask crewmembers to discuss how they were affected by each other's actions.* It is important that crewmembers understand how their actions effect each other. By openly discussing these issues, crewmembers may become more aware of the impact of their actions and the importance of communicating what they are doing and why.
- *Ask crewmembers to discuss what they were each thinking.* Encouraging crewmembers to openly articulate and discuss what they were thinking may help them understand each other's point of view and thereby enhance communication.

## A: Analysis and Evaluation of LOS Performance

For crewmembers to learn from their LOS experience, it is essential that they analyze and evaluate what happened. Crews should analyze both what went well during their LOS and what did not work as well. The analysis must go beyond simply naming the strong and weak points of their performance. Crews can gain powerful insight by analyzing why things turned out the way they did, including factors that either enabled or hindered their success. Again, the instructor must refrain from analyzing crew performance until the crew has done so themselves.

*Getting Crews to Evaluate their Performance*

It is important for crews to learn to critically evaluate their own performance in the LOS so they can carry this skill over into line operations. As the crew discusses the LOS, the instructor may find it helpful to use a wallboard to list 'strengths' and 'areas for improvement' as they are identified by the crew (Klair, 1997; chapter 3, this volume). The following techniques may be useful in encouraging crews to evaluate their performance in depth.

- ✈ *Get the crew to talk about what went well.* Discussing what they did well helps the crew to recognize what strategies were effective in managing the LOS challenges and how these strategies might be used in line situations.
- ✈ *Get the crew to talk about what could be improved, and how.* Discussing what did not go well helps the crew identify problems that occurred, examine why they occurred, and determine how to resolve or avoid similar problems in the future.
- ✈ *Troubleshooting: When the crew says everything went great.* If, when asked to evaluate their performance, the crewmembers say everything went great, facilitation may be required to encourage the crew to analyze and evaluate in more depth. If everything did go well in the LOS, encourage the crewmembers to discuss specific instances of good performance and analyze why they went as well as they did. Also, encourage them to discuss how they could have handled situations if they had not gone so well. If there were, in fact, situations that were not handled effectively, draw the crew's attention to a specific situation and ask if there is another way it could have been handled. Regardless of whether or not everything went smoothly in the LOS, it is important that the crewmembers understand both the factors that led

to their successes and the factors that led to weaker aspects of their performance.

*Eliciting Deep Analysis*

To learn deeply from the LOS experience and take the lessons learned back to the line, crewmembers need insight into why events in the LOS turned out the way they did. Crewmembers can gain this insight through in-depth analysis of their LOS performance. The instructor can help the crew analyze in depth by asking questions that require careful thought and detailed responses.

- ✈ *Ask questions that require description and analysis of LOS events.* Open questions that require descriptive or analytical responses lead crewmembers to explore issues more thoroughly, which opens the door for deeper learning.
- ✈ *Get the crew to analyze why they made the decisions they made.* Articulating *why* they did what they did helps crewmembers gain insight into their decision-making processes, as well as the factors that influenced or should have influenced their decision making. When the instructor analyzes for the crew it leaves little for them to say. It also gives them the message that the instructor is teaching them, so they are not expected to analyze for themselves.
- ✈ *Get crewmembers to discuss what they were thinking.* Getting the crew to discuss what they were thinking during the LOS can help them discover what information and events influenced their actions.
- ✈ *Encourage the crew to discuss the factors that enabled or impeded their success.* Identifying underlying factors that enabled or impeded their success in the LOS can help the crewmembers recognize similar factors when they occur on the line.

**L: Line Operations – Applying Lessons from LOS**

To help crewmembers transfer the lessons they learn in the LOS to the line, the instructor needs to encourage them to discuss how LOS performance and associated CRM issues relate to effective line operations.

*Get the Crew to Discuss Related Line Incidents*

Getting the crewmembers to discuss actual line incidents and accidents related to CRM issues that arose during the LOS can help them to appreciate the role of CRM in line operations. For example, if a crew communicated well during the LOS, which resulted in a positive outcome, having them discuss a line accident that resulted from poor communication can reinforce the importance of their effective communication.

*Get the Crew to Discuss How to Apply Their Success to Line Operations*

The next step in getting the crewmembers to transfer what they learn in the LOS to line operations is to get them to explore how they can apply techniques they utilized successfully in the LOS to overcome obstacles on the line. Getting crewmembers to talk about how they would handle difficult situations enables them to develop effective, pre-planned strategies for dealing with real situations when they occur on the line. For example, one crew's success in a particular LOS might have been aided by the fact that both crewmembers were good at explicit sharing of appropriate information. In the debriefing, the instructor might find it useful to facilitate a discussion of how to deal with crewmembers on the line who are not as good at sharing information.

*Get Crew to Discuss What They Would do Differently*

Discussing what they would do differently enables crewmembers to develop strategies they can use to make more effective decisions and avoid similar incidents in line operations. Ideally, crewmembers should discuss the strategies they can use to turn each item in their *areas for improvement* list into *strengths.* It is often useful to frame this discussion in terms of how the crewmembers can handle similar situations if they occur on the line.

*Get the Crew to Discuss How They Will Do Things Differently on the Line based on Their Experience in the LOS*

This final step requires crews to specifically tie what they have learned in the debriefing to the line. Having crews discuss how they will perform differently based on what they have learned can help them make the connection necessary to transfer thoughts into actions.

**Note**

1 For a copy of the full training manual with examples of effective and ineffective techniques in a format designed for use in the field, please contact Key Dismukes at kdismukes@mail.arc.nasa.gov or MS 262-4, Moffett Field, CA 94035-1000. Also watch for a copy in PDF format on our web page at: http://human-factors.arc.nasa.gov/IHprojects/facilitation.

## Authors' Note

This chapter is based on our study of LOFT debriefings at several U.S. airlines. The suggestions in this chapter are derived from the data from that study, our subjective impressions, the experiences the LOFT instructors shared, and general literature on facilitation. Data and references to relevant literature from the study are available in the published report: *Facilitating LOFT Debriefings: Instructor Techniques and Crew Participation* (Dismukes, McDonnell, & Jobe, 2000).

This material is presented as suggestions rather than rules because facilitation is very much a personal skill and each instructor must develop an approach with which he or she is comfortable. These suggestions provide a tool kit of techniques instructors may draw upon to develop their own style.

## Acknowledgements

This study was funded by the FAA's Office of the Chief Scientist and Technical Advisor for Human Factors (AAR-100). Eleana Edens was the program manager.

## References

Continental Airlines, Flight Operations Human Factors Group. (1992, January). *LOFT facilitation techniques: Practical strategies and techniques for LOFT facilitators*. Unpublished manuscript.

Dismukes, R. K., Jobe, K. K., & McDonnell, L. K. (1997). *LOFT Debriefings: An analysis of instructor techniques and crew participation.* (NASA Technical Memorandum 110442). Moffett Field, CA: NASA Ames Research Center.

Dismukes, R. K., McDonnell, L. K., & Jobe, K. K. (2000). Facilitating LOFT debriefings: Instructor techniques and crew participation. *International Journal of Aviation Psychology, 10* (1), 35-57.

Gibb, P. (1982, July). The facilitative trainer. *Training and Development Journal*, 14-19.

Klair, M. (1997). Briefing and debriefing techniques for instructors in aviation. In R. S. Jensen & L. A. Rakovan (Eds.), *Proceedings of the Ninth International Symposium on Aviation Psychology* (pp. 568-573). Columbus, OH: Ohio State University.

McDonnell, L. K., Jobe, K. K., & Dismukes, R. K. (1997). *Facilitating LOS debriefings: A training manual.* (NASA Technical Memorandum 112192). Moffett Field, CA: NASA Ames Research Center.

# 4 Other Applications of Facilitation in Aviation Training

GUY M. SMITH

Chapters 2 and 3 discussed facilitation of LOFT debriefing, which is the most familiar application of facilitation in aviation training. However, facilitation might also be used advantageously in several other aspects of aviation training and selection, although this is not yet common practice. This chapter explores beneficial ways to expand the use of facilitation to other sectors of airline operations. There is a caveat – facilitation is not the ideal, nor even the preferred approach for every training situation; in some circumstances traditional instructional methods are more effective. Chapter 1 discussed the relative advantages and disadvantages of facilitation and traditional lecture methods. This chapter begins with some criteria for determining whether facilitation is appropriate for a specific situation.

## Does Facilitation Fit?

Facilitation! It is not a panacea that should replace traditional methods in every training situation. Still, facilitation is not exclusive to Human Factors/CRM training. Over the years, airlines have relied on established techniques and reliable pedagogical methods to accomplish pilot training that is continually escalating in complexity, intricacy and demand. In an industry that has evolved into increased pilot responsibilities, jam-packed skies, complex air traffic control procedures, computerized systems, on-time-performance pressures, glass cockpits and fly-by-wire flight controls; does facilitation have any role in the every-day airline pilot training program? Does facilitation fit into the Systems Ground School where the object is to download the entire systems manual into a pilot's brain at an impossible pace, commonly known as the 'fire hose' teaching method. Is facilitation applicable to conducting a line proficiency check, a mandated annual assessment of airline piloting skills where the objective is to assess whether a pilot satisfies the required standards in everyday operations. Is facilitation appropriate for Captain Upgrade Training, where considerable investments into left-seat training and type ratings must be made to deliver the knowledge and skills required of airline captains? In short, does facilitation have a role in the full spectrum of airline pilot training, upgrading and qualification?

## Criteria for Using Facilitation in General Training

Acknowledging at the outset that the inappropriate use of facilitation and improper facilitation techniques may have serious drawbacks and may even produce damaging effects; it is wrong to advocate the wholesale use of facilitation as the 'only' or 'proper' way to teach airline pilots. There should be explicit criteria for determining when facilitation is the preferred teaching method. For example, it would be improper to use facilitation to introduce a new GPS system to a pilot group when Computer Based Training (CBT) lessons or a stand-up presentation, complete with graphics, models and hands-on training, would be more efficient and effective. On the other hand, it would be equally unsatisfactory to teach experienced pilots about microburst techniques without first soliciting their experiences with microburst, ascertaining the types and occurrences of microburst that they have encountered and allowing them to describe microburst incidents that they or colleagues have personally encountered.

*Criterion #1: For the proposed learning, is the collective knowledge and experience of the learners broader and more extensive than the knowledge and experience of the instructor?*

In the example of teaching a new GPS system to pilots, an instructor would waste considerable time in establishing context and 'process' when the learning objective is technical competence with the new GPS equipment and the instructor is fully equipped with the information and the training aids to train the pilots. This is a situation in which the instructor's knowledge and experience far exceeds the collective knowledge and experience of the learners, and facilitation is almost certainly not the correct teaching method. On the other hand, in the example of teaching microburst to experienced pilots, the objective is to influence behavior, particularly to develop team processes so that a crew recognizes microburst, properly analyzes the conditions favorable to microburst activity, and responds expeditiously and appropriately. Since most of the background and foundation material for the microburst lesson comes from the insights and experiences of the learners rather than from resources of the instructor, facilitation is most likely the appropriate teaching method.

*Criterion #2: Does the proposed learning have an emotional, visceral or 'ego' component that demands personal or individual ownership by the learners?*

It would be rare for a heated discussion to flare up during a training session on configuring a Flight Management System, whereas an intense debate should be expected any time that 'Procedural Non-Compliance' is taught. Systems configuration is usually a non-emotional issue for most pilots and is best taught by traditional systems training methods. In fact, in a systems configuration lesson, most pilots would view facilitation as undue tedium and a waste of time. In contrast, procedural compliance is an impassioned topic that requires a 'client-centered' approach and individual ownership of the principles of procedural compliance. Facilitation is more suitable for a class on procedural compliance; indeed, the lesson is not complete until the learners have acquired a personal attitude towards procedural compliance that will become their norm for daily operations.

*Criterion #3: For the proposed learning, is the learner relatively open-minded and receptive towards input or feedback from the instructor or are there obstacles and impediments that might limit responsiveness?*

One role of facilitation in training is to make the lesson more palatable to the learner. If there is an obstacle, whether real or perceived, teaching becomes monumentally more difficult, and the likelihood that the learning experience will fail rises sharply. Sometimes, facilitation can hurdle the stumbling block when other teaching methods are destined for failure. If a very senior 747 captain needs some coaching, the approach to training should be very different for instructors from different cohorts. A peer instructor or another veteran captain could probably be successful with a friendly over-the-shoulder lecture. However, the captain might attend guardedly or even skeptically to a discourse from a subordinate or unseasoned instructor, so a junior instructor could use a facilitation approach to make the point without seeming to challenge the experience or skill of the veteran pilot.

## Facilitation in Pilot Selection

For many years, airlines could depend on the consistency and selectivity of military experience to support their screening process. Recently, more pilot candidates originate from the civilian sector, and airlines do not have the benefit of military vetting to assist in the selection process. In selecting new airline pilots, a handful of managers, senior pilots and human resource specialists must make decisions critically important to the careers of the candidates and to the future of the airline. Many years later, when the Chief Pilot and the Flight Operations Manager contend with a non-conformist, arrogant line pilot, who is firmly entrenched in the system and supported by the pilot's union; they will wag their finger at the selection process, blaming their troubles on those who hired the misfit in the first place.

The initial phase of pilot selection is a determination of candidates' qualification, knowledge, skill and experience. Airlines have developed reliable techniques to make these assessments – written tests, simulators, physical exams, inspection of logbooks, etc.; all are established methods that do not require facilitation. However, none of these processes will reveal what an airline most wants to know about candidates: whether their personalities match the system; whether they can think critically and learn, whether they are motivated by appropriate values and goals, and whether they are capable of effective leadership. Facilitation can help to resolve these concerns. The pilot selection interview, often one hour or less, is the focal point of the selection process; here it is absolutely essential to dependably resolve these questions; a pilot's future and the prosperity of an airline rests on their accuracy. The interview process is often constrained by legal restrictions and sometimes the pilot is a skilled applicant, finely coached in 'saying what they want to hear'. With restricted time allowances and within a framework that limits the scope of the interview and allows candidates to create false impressions, the interviewers must uncover reality about the candidate's personality, learning ability, motivation, values and leadership potential. The best defenses against a bad decision in the pilot selection process are facilitation skills, particularly listening and challenging. A discerning interviewer can use a 'clarifying' process to cut through any misleading impressions that the applicant seeks to make. Very early in the interview process, skilled interviewers will divert candidates from their practiced scripts to a more realistic revelation of their skills. By hearing nuances and hints, by interpreting them correctly, and by skillful challenges, the facilitator/interviewer can clarify the issues and elucidate the truth. If an airline has only meager funds to allocate to facilitation training, those involved in pilot selection should get priority.

### *An Example*

Legal requirements have compelled many airlines to standardize their interview question bank. No matter how closely these questions are guarded, some aspects of the questions will circulate throughout the network, giving pilot candidates an opportunity to rehearse the textbook response. A skilled interviewer will listen carefully to the classic reply, select an aspect of it, and ask a follow-on question that facilitates a more individual, and perhaps more genuine answer. For example:

| | |
|---|---|
| Interviewer: | [Standardized question] How do you use CRM skills in your flying? |
| Candidate: | [Textbook Answer] To me, CRM means using all available resources when I fly. When I run into difficulties, I decide what resources are available to me and I use them all rather than just one or two. |
| Interviewer: | [Facilitation question] Could you give us an example from your flying experiences when you 'ran into difficulties' and tell us what you did? |

**Facilitation in New-Hire Pilot Orientation**

New-hire pilots typically receive several weeks of training that some airlines call 'orientation' and others 'indoctrination'. In fact, these are two distinct aspects of the training required to bring new pilots 'into the fold'. Indoctrination is a programming process that usually takes place in seminars in which the new pilot is initiated into a structured society with its traditions, standards and culture. Facilitation has only a minor role, because indoctrination is usually a one-way delivery of information. Indoctrination materials are professionally developed and skillfully packaged to expedite delivery. Videos, computer-based training, lectures and manuals are utilized to standardize the course material from class to class.

Orientation, on the other hand, is an adjustment process in which the airline builds upon skills that are already established. Some of the topics in orientation sessions are workplace ethics, sexual harassment, personal standards of conduct and, of course, the entire CRM or Human Factors awareness program. Orientation does not teach new subjects or unique material to the new-hire pilots. For example, many new-hires have already experienced at least one formal CRM program and most have already encountered crew leadership roles. During orientation, the new-hire pilot should notice a major change in the training approach, particularly a shift towards facilitation. The goal of orientation is to emphasize and highlight previous experiences that contribute value and importance to operations in the hiring airline. Since orientation further develops a pilot's experience and since the subject matter often entails perceptions and insights, facilitation should be the primary delivery mode for orientation subjects.

*An Example*

Twenty years ago, a representative new-hire class would have been predominantly male, young – ages ranging from late 20s to early 30s – and military; and a stand-up instructor taught the classes using overhead transparencies. At the end of the 1990s and into the 2000s, classes are more assorted; a wider range of material is covered; and multi –media is the typical delivery method. Currently, airline new-hire classes include more women – though they are still mostly male – a wider age span from the late 20s to early 50s, and a somewhat higher percentage of civilian over military. In an indoctrination lesson on the airline's bidding system, a stand-up instructor would teach today's class the same way it was taught 20 years ago, except a variety of resources would enhance the lecture – detailed manuals, videos, presentation slides and practice sessions. Regardless of the age, gender or background of the students, the individual requirement is the same – to understand the bidding process and to ask questions.

Twenty years ago, the new-hire curriculum would not have a lesson on sexual harassment, however, in current programs, an orientation class on sexual harassment is routine. Almost certainly, it will not succeed if it is taught in the same manner as the indoctrination lesson on bidding. It may be beneficial to introduce the topic with a lecture or a video that explains airline policy, but the actual orientation should involve a skilled facilitator – someone who knows how to get students personally involved and how to draw in the differences in age, gender and background to gain understanding of this important subject. The airline has not oriented the new pilots on sexual harassment policy until individuals in the class accept responsibility for their attitudes and behavior.

The division of new-hire training into indoctrination and orientation classifies the orientation instructors into a distinctive category. They require a discriminating selection process and considerably more facilitation training. Since new-hire orientation is so vital to a successful pilot training program, the facilitation skills of orientation instructors are major investments worthy of close scrutiny.

## Facilitation in System Ground Schools (Engineering)

The technical age swept into commercial aviation like a high plains blizzard. Starting with the posr-war era and multi-engine airliners like

Constellation and DC-6, thrusting into the jet age and Boeing 707, DC-8 and the phenomenal Boeing747, and continuing into the computer age and Airbus A-320 and Boeing 777; airline systems have never remained static, requiring constant pilot updating and retraining (Buck, 1995). New systems have always challenged instructors to find better ways of preparing and presenting increasingly complicated information to mixed classes, ranging from new-hire pilots to veterans who are upgrading to their eighth or tenth new airplane. Facilitation was a foreign word to systems instructors; their material was delivered or even dumped – usually in a bushfire approach to training (Moore & Telfer, 1997).

Before automation, systems management consisted of turning or moving knobs, switches and valves with the resultant lights that indicated the system status or response. With automation, pilots control their systems with a series of button pushes on Control Display Units (CDUs) or Flight Management Systems (FMS) with corresponding responses displayed on Electronic Flight Information Systems (EFIS) or Engine Indication and Crew Alerting Systems (EICAS) displays. Training methodologies did not maintain pace with the paradigm shift in systems. With the advent of automation came criticism of ground school training with complaints that it was too compressed and there were too many things to learn (Smallwood & Fraser, 1995). The two aircraft manufacturers, Boeing and Airbus, and the major suppliers of electronic and navigation systems differ somewhat in their automation philosophies and their approaches to systems design. Airlines also adopted slightly differing approaches to automation with distinct approaches to systems cross-checking and aircraft management (Smallwood & Fraser, 1995). The problem for the trainer is to adapt training methods to suit the learner's background, skills and experience to the present technology. Clearly, the bulk of systems training favors traditional instruction because the instructor transfers knowledge to the student. However, there may be occasions, particularly in automation classes, in which facilitation may provide a supplemental approach. For example, a systems instructor would use an uncomplicated instructional method to teach automation to automation novices, but a more facilitative approach might be appropriate for teaching automation to pilots who are switching from one highly-automated aircraft to another. For students who are familiar with automation intricacies, facilitation could draw out their experiences and help them to a deeper understanding. Training managers, who customarily excused the systems instructors from facilitation training, may want to reconsider the role of facilitation in systems ground school.

## Facilitation in Captain Upgrade Training

Except for being hired by an airline, the most momentous occasion in the professional life of a pilot is upgrading to captain. More important for the airline's training program, the knowledge that pilots must learn at this step does not primarily involve procedures or systems, and the skills that they must develop do not principally concern the traditions, standards and culture of the airline. Flying an airplane from the left seat is a matter of adjustment rather than learning. Yet, learning the skills of authority, responsibility, accountability, conflict management and leadership may require transformation of a pilot's persona (Buck, 1995). Perhaps the most significant added responsibility of a new captain is becoming a teacher. Regrettably, many captain upgrade programs do not include teaching new captains how to be instructors. Yet, every captain, willingly or reluctantly, is an instructor, particularly in times when the airline is continually hiring new pilots.

When airline pilots advance to captain upgrade training, they are already highly experienced in transport category aircraft and are old hands in the routines and traditions of the airline. More important, on every airline flight they have observed captains and have established firm notions about captaincy and about how they will operate as a captain. Having witnessed inspiring successes and dubious failures, they have vowed to imitate certain behaviors and to shun others. A thousand times they have said to themselves, 'When I am a captain, I will...'.

With a bounty of ideas and a plethora of experience, the new captains arrive at the training facility for captain upgrade training. The future of the airline may hinge on the outcome of this experience, for very few airline persons, even executives, have the potential to effect the future of an airline the way a captain can (Buck, 1995). The training events that will transform an airline pilot into an airline captain must consolidate all of an individual's experiences, conceptions and convictions into principles, values and techniques that will accommodate personality and style without compromising the standards and expectations of the airline. It seems clear that showing videotapes, lecturing and sermonizing will not accomplish the task. Sometimes captain upgrade training utilizes veteran captains to expound on habits, routines and skills that worked particularly well for them. In many cases, the training is not particularly effective because new captains have already been thoroughly exposed to role models and have already made conclusions about using these techniques. New captains should be given an opportunity to verbalize their ideas and to have them affirmed or adjusted by their instructors and their peers. With this set of circumstances, captain upgrade training is an ideal setting for facilitation.

All three of the criteria for facilitation are met: the collective knowledge and experience of the learners is broad and extensive; the learning demands personal or individual ownership by the learners; and the learner has a warehouse full of preconceived ideas that may become obstacles or limitations to their responsiveness. If the intention of captain upgrade training is to mold attitudes rather than to teach behaviours, then facilitation can acomplish the learning objectives more resourcefully than any other training method.

## Facilitation in the Chief Pilot's Office

An airline pilot once said, 'Throughout my airline career, I never want a chief pilot to know my name.' Though this sentiment is not universal, the chief pilot's office is certainly not a popular haunt for most pilots. Because chief pilots regularly deal with examination of irregularities, they often collide with personal issues, sensitivities and delicate circumstances. Often the chief pilot's job is to challenge another pilot, to demand personal accountability, or to influence a pilot to accept ownership of a situation. It is very helpful for chief pilots to consider their pilots as 'learners' and to tackle their work with the tools and techniques of an instructor - after all, their goal is to convey a lesson. Since it is common for pilots to approach the chief pilot's door with a degree of uncertainty, the chief pilot may be able to increase receptiveness by using facilitation techniques. In fact, there are few airline jobs that require more tact, understanding and diplomacy than chief pilot. Despite some pretenses of bravado, most pilots who visit the chief pilot are somewhat defensive and not totally accessible. The chief pilot may quickly surmount barriers, even some that are well concealed, by approaching the meeting as a facilitated learning session rather than as an encounter.

There are many times in the chief pilot's office, when tension is appropriate and when the chief pilot should maintain and even emphasize authority. Often, facilitation is not the appropriate technique for a chief pilot to use and it could even be counter-productive. If the situation is clear; a resolute, to-the-point discussion between two adults usually accomplishes the objective. For example, if a pilot arranges his commute too closely and misses a scheduled flight, he may have to visit the chief pilot's office. If both parties are being forthright, there is probably no need to become involved in a detailed facilitated discussion. The chief pilot can reinforce the company's expectations concerning punctuality in a five-minute manager's discussion that is more satisfactory to both pilots.

If there are uncertainties, complications, or unknowns; a more cautious and flexible approach may be proper. To illustrate, consider an example. Chief pilots are usually responsible for the performance of new-hire pilots during their first (probationary) year with the company. To assist chief pilots in their supervision of new-hires, line captains are required to submit regular evaluations of probationary pilots to the chief pilot's office. In a fictitious case, the chief pilot received several of these reports about Roland, a probationary first officer, claiming that Roland occasionally overstepped his authority, making command decisions without consulting the captain. The chief pilot decided to discuss the situation with Roland. Reviewing Roland's records, he learned that Roland, as a former US Naval Aviator, had a distinguished flying career and was a squadron commanding officer. The chief pilot's approach to Roland was one of discernment rather than one of reproach. Certainly, since he was unclear whether a reprimand was fitting, the chief pilot arranged the meeting in a small conference room rather than in his office. To reduce tension and misgivings further, the chief pilot decided to facilitate an open-minded, receptive discussion. With facilitation, the chief pilot gained Roland's confidence, and they frankly discussed his background in relation to his current situation. Through facilitation, the chief pilot was able to understand Roland's intentions and to help Roland understand more deeply what happened and why it was problematic.

There is always a danger that affable behavior by a person in authority could be interpreted as condescending or even manipulative, particularly if the pilot truly believes that an admonition is deserved. Certainly, facilitation could be used as a deception strategy to distract and mislead, however, intelligent adults quickly get the message and withdraw, dooming the gambit to failure. However, a sincere and genuine chief pilot, who conscientiously assesses the situation and uses facilitation wisely, will be believable. By equating themselves more to instructors than to policemen, chief pilots may readily gain respect and responsiveness through facilitation.

## Facilitation in Other Airline Tasks

Without a doubt, facilitation has a role in many other aspects of airline pilot selection, initial and follow-on training, and evaluating pilot performance. Instructors and evaluators who perform type ratings, initial operating experience, line checks, and proficiency checks may also have an opportunity to use facilitation as a teaching method. Facilitation is not an exclusive tool for instructors; other airline authorities may also find that facilitation can help them achieve their objectives more rapidly than direct methods. An operations manager could use facilitation to investigate daily incident reports or safety reports. A fleet manager could use facilitation to evaluate procedures. A training manager could use facilitation to develop background material for LOFT, LOE, and SPOT scenarios. A training development manager could use facilitation to review training modules, instructional models, behavioral markers, benchmarks and evaluation criteria. Instructor trainers could use facilitation in 'train the trainer' programs, practicing teaching methodologies, orientation on instructional equipment, and instructor recurrent training (Hansen & Oster, 1997). Airline management could use facilitation to develop policies and procedures, to mentor employees, to set goals, to establish management objectives, to conduct performance reviews, to conduct brainstorming sessions, to present new-hire employee orientation, etc.

In fact, facilitation has many uses, but it certainly is not a universal solution, nor should it be a fad to be replaced in a short while by another vogue. If airlines would like facilitation to become more common in their operations, there are two steps that must be carried out. First, instructors and other employees cannot be expected to use facilitation methods naturally; they must be thoroughly trained in facilitation techniques. Facilitation is more than a teaching method; it is an art. Moreover, it is an art that can be performed superbly, if it is learned well. Secondly, instructors and employees must be able to evaluate a learning circumstance to determine whether facilitation is appropriate. If the three criteria in the beginning of this chapter are thoroughly understood, the evaluation process should be easy. Facilitation is applicable to so many aspects of airline pilot training, assessment and development that it should become an essential part of everyday operations.

## References

Buck, R. N. (1995). *The pilot's burden: Flight safety and the roots of pilot error.* Ames, IA: Iowa State University Press.

Hansen, J. S., & Oster, C. V. (1997). *Taking Flight: Education and training for aviation careers*. Washington, DC: National Academy Press.

Moore, P. J., & Telfer, R. A. (1997). Learning for new technologies. In R. A. Telfer & P. J. Moore (Eds.), *Aviation training: Learners, instruction and organization* (pp. 87-96). Aldershot, England: Ashgate Publishing Ltd.

Smallwood, T. W., & Fraser, M. (1995). *The Airline Training Pilot*. Aldershot, UK: Ashgate Publishing Ltd.

# 5 Effective Crew Debriefing in Line Operations

PETER J. WOLFE

## Introduction

The success of US Airways' Altitude Awareness program, American Airline's ASAP and several other incident review processes have caused airlines to encourage their crews to debrief as a part of routine operations – not just when an irregular event occurs. Following the lead of Atlantic Coast Airlines, a number of carriers are adding 'crew debriefing' as a required action item on their post-flight checklists. Done well, such reviews improve both the performance of individuals and the effectiveness of the overall crew. As a direct result, the corporate operation is enhanced – fewer mistakes, fewer delays, greater efficiency, lower cost, etc. However, in spite of its many obvious advantages, the line debriefing still faces a challenging struggle to become an accepted aspect of U.S. air carrier operations.

Unlike most check rides, line debriefings are initiated and conducted by the crew – as opposed to being directed by an instructor or evaluator. The format, timing (where and when it takes place), duration, subject matter, scope and depth are left to the discretion of the crewmembers involved. The objective is to enable individuals on that crew to learn from the events that have occurred during their time together.

Although these reviews typically focus on situations in which two or more crewmembers were involved, that is not prerequisite. On the contrary, it is entirely appropriate for crewmembers to address any unusual, unexpected or unfamiliar situations that took place during the flight sequence – even if only one person on the crew was aware of the occurrence.

A good example of this is the preflight coordination between the Captain, Dispatcher and the ground crew when there is a need to adjust the fuel and payload in response to changing weather conditions. Of the onboard crew, the Captain may be the only person who is aware of what led up to the decisions that were made and the actions taken.

Thus, if there are lessons to be shared, it is completely appropriate for the Captain to lead that portion of the debriefing.

This chapter addresses line ops debriefings from two perspectives; the first examines why they have largely been ignored, and the second provides the principles and skills needed to make them successful. The format is intended to enable readers to compare the ideas presented here against the reality of their own operational settings.

## 'Seldom is Heard…'

The frequency and quality of debriefing varies widely. In general, U.S. airline flight crews report that debriefing seldom occurs as a part of routine line operations. As expected, flights associated with training and checking are the most thoroughly and consistently debriefed. The next most frequent debriefs occur during a new hire's first six months of probation. From then on, a self-initiated debrief is rare, unless an incident has occurred that might lead to disciplinary action or one of the crewmembers makes a special point of asking for such a review.

It is curious that debriefings are not more common. Most crews agree that the process, when done well, can be highly beneficial. In organizations where a pilot's upgrade is based on seniority (as opposed to merit or office politics), there is no risk posed by either helping other pilots improve their performance or by asking for input on ways to improve one's own knowledge and skill.

## Why Such Reluctance?

The most frequent reasons given for skipping the debriefing of a routine flight are 'they're not necessary', 'they're not welcome', and 'there's no time for them'. If true, these reasons would surely provide ample justification. But are these points valid?

'Not necessary' – This presupposes that debriefing is only justified when an irregular event has occurred or when the performance of an individual or the crew was somehow lacking. This implies that most flights are perfect and that, in fact, there'd be nothing to discuss if a debriefing were conducted. It also begs the question: isn't there something to be learned from a review of well-executed flights?

'Not welcome' – This implies that crewmembers aren't interested in or willing to take the time for a review. It is absolutely true that people have no tolerance for a poor quality debrief. However, when asked, many pilots indicate that they would willingly participate in a succinct review that highlighted aspects of the flight that caused confusion or misunderstanding, or otherwise might be improved upon.

'No time for them' – Like the other two justifications, this is valid only occasionally. This objection implies that at least one of four conditions exists: (1) All debriefings take too much time, no matter how long they are. (2) They can't be conducted while the flight or series of flights is in progress. (3) The only things worth debriefing occur during the last approach and landing before the crew is slated to quit for the day. And, finally, (4) that a crew is prohibited from conducting a post-flight debrief by a host of unyielding demands imposed by 'the system'. As with any other personal conviction, we make time for things that are important to us.

Pilots indicate that they have seldom experienced high quality debriefings. Instead they report enduring the tedium of poorly organized square-filler flight reviews or sessions that focused on assigning blame rather than identifying the lessons learned and how to do it better next time.. Given this experience, there's little wonder why the process is not well accepted.

There are four additional reasons that reinforce the reluctance to debrief. The first is our desire to be liked. We realize that the cockpit can become a tense and hostile workplace if we offend or annoy our fellow crewmembers. Typically, we are careful to avoid that risk. No one wants a reputation for being a 'junior check pilot in training' because comments made during a debrief were misunderstood or the effort degenerated into a finger-pointing session.

The second cause for reluctance arises when fellow crewmembers make it clear they're not interested in what someone else has to say – about anything! This attitude may have nothing to do with a crew's previous experience together. Instead, it may simply reflect a closed-minded outlook by one or more individuals. Instructors and check airmen in every organization tell of pilots who, upon completion of a simulator session, will say; 'Did I pass? If so, I've got to leave *now* to catch my flight home.' If that's the outlook they have toward debriefing a formal training session, there is very little likelihood that a peer debrief of line ops will be appreciated or supported.

A third cause is the realization that our own performance is not always letter perfect, completely error free or in strict compliance with standard operating procedures. Thus, we don't debrief in order to avoid being viewed as hypocrites who hold others accountable for standards we can't or don't maintain ourselves.

The final source of reluctance grows from an unwillingness to live up to the three commitments required of debrief participants:

1. The first is a commitment to 'play by the rules' of the debrief session – whatever they may be for that organization; i.e., 'Listen with an open mind. Describe events by focusing on what happened rather than who did/did not do something.' The list goes on ...
2. The second is to make the effort needed to find the 'real' answers to any questions that may be raised during the discussion and get them back to the crew – whether you're scheduled to fly together again or not. The goal is to provide credible, accurate answers to all questions raised.
3. The last commitment, and unquestionably the most difficult, is to be willing to change one's attitude and/or methods if and when a change is needed.

Our resistance to these three commitments may be the most formidable obstacle to overcome in gaining widespread acceptance of self-initiated crew debriefings of line operations. All the other reasons given may simply be a smoke screen for our unwillingness to commit our free time, to risk being proven wrong when the facts are researched, or to change the way we've become accustomed to doing things.

## Any Mentors Out There?

When trying to improve a skill or process, it's often quite useful to look at those who have a reputation for doing it well. In terms of debriefing, several flight operations come to mind: the NASA Shuttle, civilian and military flight test centers, and aerial demonstration teams. Without question, other organizations could be added to this list.

Those familiar with these types of operations attribute their success to several characteristics:

- The debrief is an expected and highly valued part of the operation.
- Participants are expected to 'check their egos at the door'.
- Comments are focused on what's right, not who's right.
- All comments are delivered in an objective, impersonal and professional manner.

- Performance is measured against established standard operating procedures (SOPs).
- All aspects of the operation are subject to review.
- No one is exempt because of position or function.
- Individuals identify what is needed to improve their personal and team performance.
- Information exchanged during the debriefing is not disclosed/ discussed outside the crew.
- Appropriate credit is given for exceptional performance.
- Brevity is prized; participants don't belabor each point.

## What Doesn't Work?

There are four traps to avoid if the debriefing process is to succeed:

1. Hypocrisy – demanding performance from others that we are unwilling to live up to ourselves.
2. Scapegoating – blaming others without accepting one's own contribution to the situation.
3. Replacing SOPs with Personal Techniques – critique based on an individual's personal technique.
4. Nitpicking and Overloading – commenting on minute details of the operation instead of restricting the review to those aspects where performance was exemplary, problematic or less than optimum.

## Setting Up a System for Success

In an ideal situation, a corporate culture encourages self-initiated debriefing at all levels of the organization. It is simply 'the way we do things around here and part of the secret to our success'. Debriefing IS an expected norm.

The concept and methods for debriefing are introduced when an individual joins the group. Formal training is provided in the skills and processes involved, not leaving understanding to chance or assuming that new crewmembers have learned the skills prior to coming aboard. The final link is forged when an individual finds that the process is applied in daily operations, as advertised, with considerable effectiveness.

Unfortunately, many organizations do not provide such an ideal setting. Thus, the responsibility for initiating a successful debriefing program falls upon crewmembers that believe in its merit. The recommendations that follow are provided by airline crews who use the process regularly and have found debriefings to be highly effective – with or without formal corporate endorsement:

*Captains:*

1. Make a personal commitment to debrief every crew you fly with, as a means of enhancing your own knowledge and performance, if not that of your fellow crewmembers.

2. During the preflight briefing of the crew:

    - Introduce the fact that you have found debriefings to be beneficial.
    - Invite input in order to improve both your capabilities as a Captain and the operation in general.
    - Alert the crew to expect a debrief of any irregular ops and at the end of your sequence together.

3. If *you* make a mistake or your performance is less than ideal – debrief it promptly.

    - It shows that you hold yourself accountable to the same standards as the rest of the crew.
    - It provides an opportunity to show how you intend the process to work.
    - Delaying the debrief implies that you may not accept critique of your performance.
    - Indicate that you feel the event should be debriefed and you'd like to open with crew input first:
        - Ask how each of them saw the situation unfold.
        - Then describe how you saw the situation and why you took the actions you did.
        - Finish by asking what lessons can be learned from the experience.

4. Although most situations don't require immediate debriefing, it is best to review/resolve problematic events as soon as possible after

their occurrence, rather than waiting until the end of the day's flying or the end of your flight sequence together. Handling events as they come up – a progressive debrief during your time together – is ideal. Many times the schedule provides ample ground time for an effective debrief between flights or an opportunity to conduct the review during cruise. Both of these occasions are quite suitable for discussion of events in which the 'lessons learned' are not likely to cause embarrassment or discord.

In dealing with more complex and/or embarrassing events, one may be confronted with a much more challenging timing dilemma. It becomes necessary to weigh the benefits and risks of conducting the review promptly vs. delaying it until a more suitable time. Debrief promptly and the issues don't fester, the matter is 'settled' and the crew can get on with its business. However, there is some risk that a serious confrontation could develop and become a terrible distraction to crew performance. Careful judgment is required of all parties – not just the Captain – and it may be appropriate to defer the debrief until after the flight is completed.

Concern for the safety of the flight is a legitimate and respected justification for temporarily delaying the debrief; 'I know we need to review what just happened, but our first responsibility is to not let ourselves become distracted while we complete this flight. Let's debrief after we get this flight to the gate and have completed our terminating checklists.'

Deferring the review, while setting a specific time to get back to it, is usually quite effective in enabling the crew to stay focused and restore the cohesion needed to maintain a safe operation.

Although it is rare, situations have occurred in which the performance of a crew is so troubled and out-of-synch that it demands immediate attention if the flight is to conclude safely. In such circumstances, it may be necessary to temporarily stop the flow of events. On the ground, one can always set the parking brake or delay the takeoff until the issue is suitably resolved. Enroute, although the options are more limited, there is (nearly) always an option to ask for holding, go around, execute a missed approach – do whatever it takes to insure that the crew and aircraft are brought back in-synch and re-focused before proceeding.

5. The following three questions have been found to be quite useful in helping the Captain create a positive atmosphere for an end-of-tour debrief:

- 'Were there any aspects of our operation that were unexpected, confusing or uncomfortable?'
- 'Do you have any recommendations that we should pass along to the company in order to improve the operation in some way (safety, service, training, efficiency, etc.)?'
- 'Do you have any suggestions for me – ways I might improve my performance as a Captain?'

A quick scan of these recommendations reveals the fundamental difference between a traditional line check debriefing versus the more informal, interactive exchanges that occur when using this approach. In line operations, self-critique is the key to debriefing success. Most pilots are much harder on themselves than any peer would dare to be. This process takes maximum advantage of that attribute – usually obtaining candid and insightful observations about what occurred during a flight.

This approach enables crewmembers to keep their egos intact. It also affords the opportunity to debrief while enroute without increasing the risk that the discussion will become a distraction or disrupt the crew's ability to work effectively together. By conducting the debriefing in this manner, the crew does not allow a problem to fester or to distract them for any longer than necessary. In addition, crew cohesion is restored, if not improved, and you haven't added a minute to the duty day.

### *First Officers and Flight Engineers*

Most of the recommendations by and for Captains can be employed successfully by other crewmembers as well. Even if the Captain is not proactive about debriefing, a subordinate crewmember can initiate one – without jeopardizing the professional respect of fellow crewmembers.

If an irregular event occurs, ask to review it so that you might learn as much as possible from the experience. If you feel that your own performance wasn't up to par, again, ask for some feedback from those who are more experienced with the operation. Invite their input on how you might have handled the situation in a more effective manner.

Even if the flight sequence was completely routine and nothing irregular occurred, at least an end-of-tour debrief should be conducted to insure that each individual's skills are kept as sharp as possible. The Captain is not the only person who can initiate such a review – anyone on the crew can do so. A great many pilots indicate that they would like feedback on their performance but are reluctant to ask for it out of concern that they may be viewed as a 'goody two shoes'; they feel they need a special reason to justify the debrief.

There are a number of good reasons one might give, for example, 'I have a checkride coming up shortly', 'I'm new in this crew position and haven't seen this sort of operation (under these conditions) recently', 'I'm going to be upgrading soon and want to be as prepared as possible', etc.

If, in turn, the Captain or others on your crew invite your input on an end-of-tour debrief, respect the opportunity and give careful thought to your comments.

*General Recommendations for All Crewmembers*

As the flight sequence evolves, each crewmember should make mental or written notes of the items they've observed and might want to address during an end-of-tour review. Once again, it is important to note that the list should include items that were particularly effective as well as those that were not.

Unlike the line check, in which the evaluation is based on a snapshot view of one's capabilities during a single flight, routine operations afford our fellow crewmembers a much better chance to see a more accurate representation of our knowledge and skill over the course of several flights. Unfortunately, check airmen are forced to assume that any error or shortfall they observe is typical of performance and not an anomaly – and they often debrief each item accordingly.

Line crews, on the other hand, should avoid debriefing each and every lapse or error. Instead, it is important that they carefully consider two questions before deciding to mention an item in a crew-initiated debrief: (1) Does the item involve a compromise in safety of flight or standard operating procedures, and (2) Is the item being debriefed a trend or is it an isolated occurrence? Items that clearly compromise safety or SOPs, even though they may be one-time occurences, should be debriefed as soon after the episode as possible. Items that are within SOPs but represent threats to effective crew performance or operational efficiency are excellent candidates for the end-of-tour debrief. Of these latter items, those that are repeated occurrences are ideal points to raise in this sort of summary review.

We've probably all been counseled at one time or another about the importance of balancing our discretion against our valor when it comes to handling potentially explosive situations. Crew debriefings are a perfect example of such a dilemma. Some individuals are so closed-minded that they simply will not listen to any form of critique, no matter how valid it may be. In that case, we are faced with the challenge of determining whether there is anything to be gained by presenting an uninvited and unwelcome debrief 'for the record'.

When this question is posed to most Professional Standards committee members, their response is typically; 'If the issue bothers you to the point that your performance is degraded or you would consider reporting the individual's behavior/airmanship to the Chief Pilot, then it is sufficiently serious that you should address the matter with that crewmember directly. We owe this much to each other as a mark of our professional respect for fellow aviators.'

An equally challenging situation arises when we are invited to offer input, but we are either unfamiliar with the attitudes of our fellow crewmembers toward objective critique or we've assessed by their prior actions that they don't want to hear anything negative about their performance. In either case, it may be best to limit our recommendations to just one or two 'high order' items – supported by at least one solid example from your experience together – that you feel is/are in need of attention. Handling the invitation in this way, you not only maintained your integrity and offer some constructive information, but you present the idea in a manner that is most likely to be heard and understood.

For example, let us assume that you observe that a Captain (1) is inconsistent in using the checklist, (2) is sloppy about making company-prescribed callouts during takeoffs and landings, and (3) pays little attention to sterile cockpit procedures.

Reviewing your list of observations, you can readily group the items under a single high order principle; in this case, compliance with SOPs. Thus, if the Captain asks for your input, you could offer the following observation, 'You are more relaxed about the SOPs than most of the pilots I've flown with here, and that makes it difficult for me to know just what to expect as we go along. For example...' Other high order principles would include communication, situation awareness, teamwork, workload management, etc.

Why does this work? Because most of us recognize that our performance isn't perfect and can certainly be improved in some way. However, even the most open-minded individual may find it difficult to accept a lengthy laundry list of self-improvement tips from fellow crewmembers.

When asked, most line Captains report that they seldom receive feedback on their performance from their fellow crewmembers. They frequently wonder how they compare with other Captains and whether there are ways in which their performance might be improved. It hasn't been the custom to ask subordinate crewmembers for feedback, thus it has not become a common occurrence on the line. However, those who do invite the critique have found that the insight gained has proven to be extremely beneficial to both their technical knowledge and skills and their flight/crew management capabilities.

Whenever one offers such observations, it is vitally important to recognize and accept (1) that our own performance may have made the situation unfold as it did or worsened it, and (2) that others' perspective and experience may have enabled them to view the situation differently and, perhaps, or more accurately.

## Summary

Although crew initiated debriefings of line operations are not a common occurrence, there is considerable evidence that most crewmembers would invite and appreciate this form of feedback. Like any other skill, highly effective debriefing takes practice. Ideally, the corporate culture offers both guidelines and training for crewmembers on how to conduct such reviews, and reinforces the practice through a variety of measures.

The recommendations presented here reflect a cross-section of ideas that have been found to be effective – whether or not there is corporate-level support for debriefings. Without question, the greater the support for this effort the better, but it can flourish based solely on the initiative and professionalism of individual crewmembers.

As line crewmembers, we typically observe each other over a sequence of several flights. For this reason, our insights and assessments are potentially the most accurate and comprehensive of all. Clearly, this is the most valuable learning opportunity available to any crewmember. It's a resource that should not be ignored or discounted. Although individual debriefing styles may vary, the bottom line issue should always be: Did each crewmember have an opportunity to help the rest of the crew learn from and improve upon their performance as a result of our experience together?

# 6 The Mediated Debrief of Problem Flights

MARK B. KLAIR

## Introduction

When an airliner crashes, the National Transportation Safety Board (NTSB) sends a team of investigators to determine the cause of the accident. The highly qualified team of professionals arrives at the scene and works with members of all parties with a vested interest in the accident to determine the cause. The investigation is meticulous, extremely thorough, and may take years to complete. The objective for the NTSB is to prevent the reoccurrence of similar accidents, especially those involving loss of life and destruction of property. The NTSB publishes a detailed public report describing the facts of the accident, the analysis of those facts to find the root causes, and recommendations for preventive measures by the industry and the Federal Aviation Administration (FAA). Lessons learned from NTSB reports are invaluable training tools for airlines. Fortunately, major airline accidents are infrequent; thus NTSB reports provide only a limited sample of crew performance and safety issues.

Lessons can also be learned from flights that do not end in accidents. For example, the Aviation Safety Reporting System (ASRS) was developed by the National Aeronautics and Space Administration (NASA) with FAA input to enable pilots to report unsafe occurrences that take place in the cockpit. The ASRS has developed a large, valuable database available to anyone concerned with the problems pilots experience. This database has been used extensively in the aviation industry to identify problems and potential hazards.

Although extremely valuable, the ASRS database has distinct limitations. It contains extensive examples of what happened in problematic flights, but typically provides little information about why those problematic events occurred. For example, the database describes large numbers of altitude deviations, but provides limited insight to why so many deviations occur. In ASRS reports, crews frequently describe their own mistakes, but are seldom able to provide more than a superficial analysis of why they made those mistakes. A pilot might report mistakes the crew made while performing an approach and confess that the crewmembers were not communicating well that day, but probably would not be able to analyze in depth the sequence of events underlying the poor communication and contributing to the incident. Yet understanding the deep causes of crew error is essential if aviation safety is to be improved.

Another limitation of ASRS reports is that they typically focus on what went wrong in a flight and seldom identify the things that went right. Yet a deep analysis requires putting what went wrong into the larger context of all the things the crew did well. Also, airline managers, pilots, and pilot unions need information about the performance of their own crews, and this information is not available from ASRS.

To complement the information available from NTSB and ASRS reports, airlines need another type of analysis that will enable them to:

1. Find the root cause/causes of incidents.
2. Help the flight crew learn from the experience.
3. Address any issues the crew may have with procedures, training, or personal conflicts with each other.
4. Help the *entire* crewforce learn from the experience by disseminating lessons learned without revealing the identities of the incident crew.

To meet these objectives I developed the Mediated Debrief as a specific, structured procedure and program to be administered by airlines, pilots' unions, and other organizations. The Mediated Debrief provides an in-depth analysis of incident flights by assisting the crewmembers in conducting a debriefing. The debriefing is mediated by a facilitator, who helps the crewmembers address uncomfortable issues and resolve personal conflicts. Facilitation is crucial; it can resolve conflicts and promote learning in difficult circumstances in which traditional methods typically fail.

The Mediated Debrief has two main parts:

- Crewmembers' recollections of what happened on the incident flight. Starting with the most junior person, each crewmember describes the flight in chronological order, beginning with their meeting in the flight operations area before entering the airplane.
- A simple analysis by each crewmember of what he or she did well during the flight as what he or she would want to improve. As each crewmember talks, the debriefer writes this analysis on a noteboard using the words of the crewmember as much as possible.

After the crew's analysis is completed the debriefer recaps the lessons learned, answers any questions, and addresses any unresolved personal conflicts. After the debriefing is over, the debriefer writes a summary of the lessons learned (de-identifying the participants). Mediated Debrief summaries provide information valuable to the entire crewforce and they are useful training tools that can greatly increase the safety of flight operations.

## What Triggers a Mediated Debrief?

Mediated Debriefing is useful in any situation in which participants in a significant event or incident would benefit by discussing what happened and why it happened. The process may be triggered in various ways. The airline might receive a Letter of Investigation from the FAA concerning a possible violation of flight regulations. Problematic flights may be brought to the attention of the airline's safety department or the safety committee of the pilots' union. Crewmembers themselves may request a debriefing. For example, in one case a crew submitted a safety report after making a night visual approach into an area of high terrain and having to perform a Ground Proximity Warning System (GPWS) escape maneuver. In other cases crews have requested debriefings to resolve personal conflicts.

The conduct of the Mediated Debrief is kept separate from any disciplinary action or retraining required of the crew. Management may conduct its own investigation after an incident, working with information provided by the safety department. This investigation may lead to providing additional training to the crew, if needed, or – if reckless or purposeful misbehavior occurred – disciplinary action. However, in many situations additional training and disciplinary action are neither appropriate nor useful for improving safety. In these situations the Mediated Debrief often provides a better approach to analyzing events and helping a crew learn how to prevent similar occurrences. The Mediated Debrief also has the enormous advantage of providing a way to disseminate lessons learned to the entire crewforce. The FAA has looked favorably on the Mediated Debrief program because it shows a good faith effort by the airline and its crews to investigate and correct errors and to benefit from the knowledge gained.

## An Illustration of the Need for Mediated Debriefing

Here is an example of an actual airline incident in which poor team building led to an off course deviation of eighteen miles. While cruising at altitude, the wide-body aircraft suddenly turned 30 degrees to the right. Before the crew could return the aircraft to the planned flight path, they had flown eighteen miles off course. Although no other aircraft was in the flight path of the aircraft, the excursion off the designated flight route was a violation of the Federal Aviation Regulations (FARs). The crew was sent a Letter of Investigation from the FAA, the first step towards certificate action for a possible flight violation.

Traditionally, when an aircrew is involved in this sort of incident, they are expected to write a statement to the FAA. The FAA would then expect the airline to provide extra training, in the form of simulator and 'line check' evaluation, to ensure learning has taken place to prevent any future repetition of the violation or incident. The simulator is used to evaluate the crew and usually includes normal and abnormal procedures, with emphasis on the technical procedures that may have led to the flight violation. Crewmembers are usually scheduled for these events as soon as possible to allow them to get back to the line quickly, but are not necessarily scheduled to fly the simulator together. In fact, not all of the crewmembers may do the simulator training or receive the line check. For these reasons, the crew may not discuss the flight in enough detail to find the root cause or causes, and to distinguish what they did poorly from what they did well.

In a case such as this, prior to Mediated Debriefs, the crew may have spoken to flight management, the safety department, and to the FAA and explained what they were doing just prior to the aircraft turning to the right. They might have explained how the aircraft made a turn and strayed 18 miles off course before any corrective action was taken. When asked what made the aircraft turn, they might have said that there had been some confusion as to the navigation system being set up for Inertial or radio navigation. They would have been cautioned to be more attentive to detail in the future to ensure quicker corrective action of the flight path. They would receive an evaluation in the simulator and a line check evaluation in the aircraft.

In this case, using the Mediated Debrief, we were able to learn exactly what caused the inattention to detail. We found out why crew coordination was poor, which prevented them from responding more rapidly to correct the deviation.

As would be expected, during the Mediated Debrief, when the crew started to discuss the flight event, they focused immediately on the unexpected deviation. Since this was the problem that caused the alleged violation, this is where most people would feel the need to start. However, following the Mediated Debrief format, we began the debriefing with how the crew first met. The crew reflected back on how they met and individually explained the events leading to departure. After the first officer (FO) and captain (CA) met, the FO decided to stop to check e-mail and did not accompany the CA out to the aircraft. At this particular company, the CA and FO normally ride out to the aircraft together in a shuttle that takes about 10 minutes. This time is usually used informally to establish the team concept. Team formation is essential in creating an atmosphere of free communication among crewmembers. This communication is essential for allowing the CA to make informed decisions during the flight. In this case, there was a slight resentment by the CA towards the FO, although this would normally not be considered much reason to prevent communication between two individuals. The CA did not think much of this until the second officer (SO) asked for and commented on the whereabouts of the FO.

The resentment increased as the CA started to think about this until the FO arrived at the aircraft. There was just enough tension in the cockpit that the CA did not brief the FO on how he wanted the navigation set up for the flight. The FO recalled there was some tension in the cockpit, but was unaware of the reason. He also felt uncomfortable enough not to ask for any clarification as to the CA's expectations of his duties.

This is an important point, which we have learned from the Mediated Debriefs. Most people may think of a miscommunication between individuals as a result of a very emotional exchange or an angry dispute that takes place. In fact, it takes very little tension between people to make someone uncomfortable enough not to challenge while monitoring, or even ask a question for clarification. Most of the issues that spawn personal conflict are so insignificant that it takes some effort to elicit them from crewmembers in the Mediated Debrief. Much of the time when they come to realize what 'started it all', they end up laughing about it. This is when I know progress has been made in the learning process, as well as in the conflict resolution. Though the FO arrived at the aircraft 'on time' legally, the CA and the SO perceived him as being late, creating more unspoken resentment that affected the communication process during the pre-flight briefing. Since the expectation for most crewmembers is to ride out to the aircraft together, the FO did arrive much later than normally expected. Because of this breakdown in the crew formation process, expectations of the CA regarding navigational procedures and duties were not verbalized to the FO. The CA had expected to fly using VOR, ground-based radio navigation to navigate from one point to the next, because the Inertial Navigation System (INS), which uses the internal aircraft navigation system, was not set up properly. In his rush, the CA decided to omit the updating procedure required by their organization once confidence in the

INS is lost. Instead, he elected to use the VOR system for radio navigation. However, the FO was not made aware of this and continued to fly using INS rather than VOR. As the aircraft came to a waypoint between VORs, it turned to the next point. Confusion ensued as differing VOR and inaccurate INS positions led the crew off course.

In this particular Mediated Debrief, the company's subject matter expert (SME) for the aircraft was present along with the debriefer. The SME discussed technical aspects of the navigation system and standard operating procedures (SOPs). As a result, the group was able to identify pre-flight procedures the crew had performed in a nonstandard way, and was also able to reinforce positively the crew's correct use of other procedures.

With the Mediated Debrief, the lessons learned were far more thorough and more beneficial – not just for the crew involved, but for the entire crewforce – because the lessons learned were used later in recurrent training for all the pilots at that airline. The lessons were about professionalism, team formation, crew debriefing, communication of expectations, how to prevent and resolve personal conflict, crosschecking, monitoring, and navigational procedures, to name a few.

How did the above lessons evolve through the course of this Mediated Debrief? The entire error chain only came to light during the Mediated Debrief. The crew examined their performance from the time they met to the landing and shutdown. They discussed the need to address the tension that grew within the cockpit, and to keep the communication flowing. The crew was led through a discussion about the dynamics involved with tension and its effect on communication. They came to the conclusion that simply addressing the tension in a non-offensive manner would have helped reestablish an atmosphere of effective communication. Using standard Mediated Debrief procedures, the debriefer first asked the SO what he had done well. The SO focused on his technical knowledge and used checklists, which is typical of second officers. The other crewmembers stated that the SO had performed his duties well. The SO noted, however, that there was something he could have done to help prevent the problem. Even though it would have felt unnatural or uncomfortable, he could have addressed the tension or at least responded through his communications in a way that would have reduced the tension. Specifically, he could have expressed the resentment he shared with the CA in order to communicate effectively with the FO.

During the Mediated Debrief the FO was made aware by the other two crewmembers of the reason for their slight resentment. This was not the first time he had started the flight by checking e-mail and arriving at the aircraft later than the CA. Though he arrived at the aircraft on time according to the company's operations manual, he was *perceived* to be late by the crew. This created unspoken resentment, which affected the communication process during the pre-flight briefing.

The debriefer asked the FO what he could have done to improve his performance. To start with, the FO said, he would act more professionally and leave the e-mail for another time. He understood his responsibility in team building. He realized that it was important to work with the CA to establish the team. Another important lesson learned was the need to recognize when other crewmembers weren't responding to him in a normal manner. The debriefer continued to press the FO for specifics of what he noticed in the way the other two crewmembers were communicating with him. He knew 'something was not right', but did not know if he had done something to upset the crew – he just sensed tension in the cockpit. Like the SO, the FO realized during the debriefing the need to address that tension to keep the communication lines open.

The CA initially stated that there was not much he had done well during the flight. He said they would not have been in the Mediated Debrief if he had. When the crew had entered the Mediated Debrief they were understandably upset. However, by this point, the tension among them had decreased due to the format of the Mediated Debrief. As a result, his crew was now quick to point out that he had done a lot of things very well. They remarked that he flew a 'good airplane' and worked well with the crew. This remark was interesting in that it came from the FO himself, even though he was the individual with whom the CA had the personal conflict.

The format of the Mediated Debrief facilitates this sort of positive feedback, as will be described in detail later in this paper. The debriefer starts by asking each crewmember what he or she did well. If a crewmember has difficulty answering that question, the facilitator asks the other crewmembers to provide the answer. The very nature of this format aids in the resolving of personal differences.

In the above debriefing, reducing the tension seemed to enable the CA to state more emphatically what he could have done to improve his performance. This included stating more clearly his expectation of the crew. The CA stated that he should have explained to the FO that he would rather the FO delay the e-mail for a later time. The FO chimed in at this point that the CA should not have needed to do this. The CA continued, stating that his debriefing was non-standard in that he was not thorough enough, and this was because of the tension. The CA stated that since he is ultimately responsible for the aircraft and crew by FAR 91.3A, he should have been the one to address this tension. The other two crewmembers responded that they were all responsible for the communication during the flight.

The learning for all the crewmembers increased even further when the crew discussed the specifics of how to recognize when tension exists in others, even when you don't happen to feel the tension yourself. The crew discussed the fact that social distance can be sensed from body language, intonation of voice, and hesitation to communicate or establish eye contact.

Why was this discussion important? The crewmembers are there to learn whatever they can from each other to prevent the incident from recurring and to reinforce what went very well. The time spent discussing the specifics helps the crewmembers retain the information and see the importance of the points – all part of the experiential learning designed into the Mediated Debrief. The CA continued to explain, with the help of the other crewmembers, three other items specific to the aircraft SOPs that he neglected to complete. While they had no bearing on the incident being debriefed, these were things that he could improve. The representative from the technical standards department (acting as the SME) was able to reinforce the SOPs the CA had missed and answer other related questions asked by the crew. Once the crew had established a better working relationship in the debriefing, they were able to determine why they failed to re-establish the correct heading immediately after the aircraft made the unexpected turn. The main reason for this was not keeping up with the route of flight enough to know what the proper heading should have been. At this point, the standardization representative reinforced the correct company procedures for enroute navigation for the crew.

If, while reading this, you are starting to shake your head thinking this is a painfully in-depth analysis of a small event, you are not alone. I would have thought the same before conducting quite a few Mediated Debriefs. In this example, the error involved was one of omission of the CA's expectations of how the FO should set up the navigation, as well as situational awareness of the flight path once airborne. However, it takes only a very small mistake to cause a major deviation from an intended aircraft flight path. In my experience the overwhelming majority of crewmembers have been pleased to be able to discuss whatever they could to ensure they would not repeat the incident or event in their careers.

Small mistakes can lead to loss of life in fields such as medicine and law enforcement, just as in aviation. Thus, it is essential that teams working in critical occupations concentrate on setting and maintaining an atmosphere of open communication. This is a very important point to stress to all members of *any* team, and I do it every chance I can when I see personal differences arise between people. The healthy egos involved in any critical occupation tend to be a great obstacle to communication. Even when people realize tension is preventing communication, they do not know how to reestablish communication and often do not want to do so. When egos are involved, emotions can quickly prevent rational thought, which is needed to discuss critical operations. This, in turn, makes it difficult to reestablish communication, and hence *little* or *no* communication takes place. Thus, addressing the tension is much easier said then done. Consequently, it is extremely important to establish the team concept and atmosphere when the team first comes together. Having done that, it is easier to keep the lines of communication open and address differences in expectations as they arise. People will communicate with each other more effectively if they have a mutual respect for each other. With the many preconceptions or filters we communicate with, it is easy to misunderstand what someone said. **When the proper team concept has been established, crewmembers are more likely to interpret ambiguous words or actions in a positive way than in a negative way.**

## Structure and Process of the Mediated Debrief

This section describes in detail how to organize and conduct a Mediated Debrief. A later section provides two examples of using this process taken from debriefings of actual flights.

### *The Players*

In debriefings of flight operations, the participants in the Mediated Debrief typically include:

- *The Manager of the Safety Department* or his representative, to ensure that the program runs as expected. A representative of the pilot union's safety committee may also attend.
- *The Debriefer*, who leads the session and facilitates the participation of the other members of the group.
- *A Subject Matter Expert (SME)* to answer any technical questions.[1]
- *The Entire Crew*. Flight managers have sometimes attempted to arrange a Mediated Debrief without all crewmembers present because of logistic or cost problems. Partial-crew debriefings are much less effective and should not be attempted. Nor is it useful to attempt to conduct the debriefing by telephone, as one organization attempted with poor results.

Some unions are developing their own Mediated Debrief programs and will most likely offer to conduct most – if not all – Mediated Debriefs that deal with personal conflict. This approach may help make the pilots feel more comfortable by handling the matter outside of the airline's management structure.

*The Setup*

Since the Mediated Debrief relies almost completely on the recollection of the crew, it should take place as soon as possible after the flight or series of flights in which the problem occurred. The FAA certainly would expect any company to respond in a rapid and positive manner to address any problems that have occurred. Ideally, the debriefer can meet with the crew and SME shortly after the problem flight. This is not to say that a Mediated Debrief should be conducted while crewmembers are still in a highly emotional state. However, in the case of personal conflicts in which crews have exchanged emotional comments, a long delay between the incident and the debriefing allows defense mechanisms to drive wedges between crewmember perceptions and the reality of the flight. Mutual respect erodes. If a longer delay is unavoidable, it may still be possible to resolve long-standing personal conflicts after longer delays, and it is worth trying because personal conflicts can endanger safe operations. Most organizations set an arbitrary limit of two weeks between an incident and debriefing to provide the best chance for success.[2]

A great part of the success of the Mediated Debrief depends on the attitude of the crew as they first walk through the door. The method of notification of the crew of the Mediated Debrief is important. The person who notifies the crew should not be from the department responsible for discipline. In most airlines the safety department is appropriate. If the debriefing is sponsored by the union, notification should come from the professional standards or safety committee. The tone of the notification should not be accusatory. Whoever organizes the debriefing must ensure that all of the players are available before notifying the crew; more than one Mediated Debrief has been canceled due to a lack of planning and coordination among the parties involved. The notification should include the time, place, and objectives of the Mediated Debrief.

If the players who are not crewmembers have no previous experience with Mediated Debriefs, they should arrive early so the debriefer can explain their roles and exactly what is expected of them. Some information may be passed to them about the overall situation to aid in organizing thoughts prior to starting the Mediated Debrief, but it should be clear that no hearsay should be used then or in the Debrief itself. I have been asked whether it is a good idea to use prior written statements from the crewmembers. This depends on the actual content and the relevance to the flight, but generally it is not a good idea. Prior knowledge of an incident or the relationship of the crewmembers can be useful if the information is reliable. Often, however, participants have written personal statements while highly emotional; if these statements are introduced into the Mediated Debrief, participants may become emotionally entrenched. I normally will not read anything or talk to any of the crew prior to the Debrief in order to avoid giving the impression of having chosen sides.

The key to a successful Mediated Debrief is making crewmembers as comfortable as possible in an uncomfortable setting. I always ensure that the seating is random so the crew does not feel as though they are walking into a formal inquiry, sitting together on one end of a table across from the 'other guys'.

*Room Equipment*

*Table and Chairs.* Before the Mediated Debrief begins, ensure that there are enough chairs around the table for all of the participants who will be attending.

*Technical Manuals.* Include whatever manuals or charts the SME may require to discuss flight events, technical standards, and company policies. This helps reinforce the crew's awareness of company technical procedures and practices.

*Human Factors/Crew Resource Management (CRM) Manuals.* Include the company's materials on CRM or human factors procedures; this helps reinforce the crew's awareness that the company has specific policies and procedures for CRM/human factors.

*Whiteboard/Chalkboard.* The room should have a large whiteboard or chalkboard for listing what went well in the flight and what could be improved (as discussed in the next section). Ensure that enough fresh markers or chalk are available; nothing is more disconcerting than running out of ink or chalk in the middle of a stressful recounting of a near death experience.

*Pads of Paper.* All of the participants may want to take notes. As a debriefer, I take notes to keep the debriefing organized and to prevent rambling, which is lethal. I also use these notes after the session to prepare a de-identified, lessons-learned summary that the organization can use for training.

*Key Issues to Remember*

When conducting the Mediated Debrief, the following issues should be considered:

1. The pace should never be too slow or labored. A brisk pace in the less than significant areas will help make the participants more comfortable. The goal is a fair and accurate representation of crew performance during the flight, and what they would do differently (if anything) if the situation occurred again.
2. The tone set by the debriefer should be one of an objective recorder, devoid of judgement (unless a crewmember tries to take control of the debriefing, or exhibits a misunderstanding that might undermine the debriefing). For example, some defensive crewmembers have characterized their emotional outbursts during the flight as 'attempting to communicate their expectations to the other crewmembers'. They claimed that their human factors or CRM departments taught them to do this. In this situation it was necessary to let these individuals know they were incorrect – in fact their attitudes were the antithesis of what their company had taught them. This is rare, however, and the overwhelming majority of crews are attentive and cooperative once the tone of the Mediated Debrief is properly set.
3. Cohesive team building is an important goal, and every communication by the debriefer must focus on it. At the completion of a Mediated Debrief involving personal conflict, the debriefer may ask the crew if there would be any reason they would not fly together again. Alternatively, the debriefer might ask what they would say to each other to ensure a safe and uneventful flight if they received a call to fly together again the next day.
4. When eliciting information for discussion, the debriefer's priority should be:

   - *The Crewmember* whose turn it is to speak
   - *Other Crewmembers,*[3] after the crewmember speaking is done
   - *The Debriefer and SME*, if the crewmembers fail to discuss critical issues.

5. Whenever appropriate, use a sense of humor. If you do not have one, buy one. It can be essential.
6. Whenever possible, relate the lessons learned to the prior training received by the crewmembers within the company, both for human

factors and technical procedures: this helps link training and operations.

*Procedures*

The Mediated Debrief is conducted in four stages: *Introductions, Narrative* of 'what happened' by each crewmember, *T-list* of what went well and what could be improved, and *Recap* of the T-list by the debriefer.

*Introductions.* When the crewmembers walk into the room, introduce them to the other players. Once everyone is seated, concisely introduce the crew to the process of the Mediated Debrief. They need to know the role of each person sitting with them, the structure of the Mediated Debrief, the objectives, and whether or not the company will use information from the debriefing to inform others in the organization. They also should be reassured that the debriefer will spend as much time as they want or need to ensure that their concerns about the flight and post-flight events are voiced, and that any questions will be answered. If the crew is not familiar with the Mediated Debrief program, they will be curious about the process, how it is used, and who will be privy to the facts gained from the debriefing.

*Narrative.* The debriefer will encourage each crewmember, in turn, to talk about what happened in their own words, without interruption. The only interruptions should come from the Debriefer or subject matter expert for clarification purposes. It is especially crucial to control interruptions in highly emotional situations.

The junior-ranking crewmember speaks first. It is dangerous for the CA to go first because the CA sets the tone in any team, and this may make it difficult for other crewmembers to express uncomfortable recollections or opinions in a controlled environment. This happens whether or not the CA intends this influence. By the nature of the crew hierarchy, crewmembers experience power distance, great or small.

The debriefer guides the crew to begin the discussion with their first meeting and to describe the first leg in some detail. The crew should discuss the flight phases, including the *briefing* conducted by the CA, the *start*, the *taxi* and the *take off.* From there it is entirely up to the debriefer to decide which parts of the flight or series of flights are relevant and should be examined. To determine this, the debriefer asks the crew how many legs were flown and which ones were 'significant'. [4]

Once the junior crewmember has finished, the debriefer cycles through the rest of the crew, letting the CA have the last say. Differences in recollection among the crewmembers may occur. The debriefer should try to resolve these differences, but only if they are critical to the event.

*T-List: The Board Work.* The debriefer draws a large 'T' on the board creating two columns with *Well* on the top left side of the T and *Improve* on the top right side of the T (see Figure 6.1).

| | WELL | IMPROVE |
|---|---|---|
| SO | | |
| FO | | |
| CA | | |

**Figure 6.1 Sample T-List**

The debriefer should write what each crewmember says on the board without editorializing, other than for clarification. I normally write the exact quote of the crewmember on the board. The *Well* and *Improve* sections should be specific and detailed, but concise enough to be meaningful to the crew and others who will learn from their experience. For example, writing the word 'assertion' may not be enough to point out how a crewmember might have stated his or her concern, problem, and solution during a critical maneuver.

Starting with the junior crewmember, and ending with the CA, ask each crewmember the following two questions:

*1. What did he or she do well during the first significant phase of flight?* Depending on the company, the team formation might be significant, so how they introduced themselves would be relevant. In the case of Shuttle crewmembers from NASA, who have been working for months together as a team prior to flight, the formation phase is not as relevant. The same goes for crews from smaller airlines and military squadrons who have flown together for years.

The focus should be on what specifically the crewmember did well. The initial response from most crewmembers is typically 'not much'. Aviators and competitive team members of any critical occupation are rarely given a chance to discuss their strengths in a situation in which they feel uncomfortable unless in defense of actions. This can work well for the debriefer when trying to foster an improved team environment among the crewmembers.

If crewmembers are reluctant to speak glowingly of themselves, an effective technique is to ask one crewmember to describe what another crewmember did well to support the CA during the flight. This does several things for the team. First, it lends credibility to the debriefer. Most of us tire of individuals blessed with the ability to point out all of our mistakes, of which we are already well aware. We do not tire of those who point out what we have done well, however. Socially, it is more acceptable to have others praise us than to do so ourselves. This makes us more comfortable, and whoever is responsible for that gains the credibility in our eyes to be able to critique our performance. Among crewmembers this fosters an atmosphere conducive to resolving personal conflict. If a crewmember does say a great deal about his or her own performance, then it probably is not necessary to ask others to expand on it. That individual is already feeling confident about his or her performance.

*2. What, if anything, could the crewmember improve given the same circumstances again?* Unlike the *Well* side, it is not normally good for the other crewmembers to comment on this. Most crewmembers will state readily what they could improve. If not, the debriefer can attempt to bring out the information by asking questions about their actions in the incident phase of flight. It is very rare that one would need to go much further than a simple prodding as to what the crewmember might do to avoid the situation again. If this fails, however, the SME should be brought into the discussion since it is the debriefer's duty to ensure the crewmember does not leave the room without understanding what he or she did well or needs to improve upon. The SME should be asked to comment on what the SOP or acceptable behavior is for that phase of flight. Once covered, the *Improve* side will then include what the SME brought up with the concurrence of the crewmember. I have yet to see any situation in which the crewmember disagreed with the SME or debriefer on what needed improvement.

The discussion becomes more detailed as each crewmember adds to what the others have said. I ensure that each crewmember is reinforced by the others in the lesson learned. For instance, one crewmember hinted to the CA that there was terrain to the left and right as they descended on a night visual approach. During the Mediated Debrief he learned that the CA had been listening to him through filters of fatigue and complacency. This crewmember should have followed his airline's standardized procedures for communicating operational conflict, clearly stating his concern, define the problem, and the solution. I then asked the CA if he really thought that the standardized procedure would have helped him make a different decision. He explained to his crewmember and us that if they had been flying during the day, when he was wide awake, or on an unfamiliar approach, the hint would have been enough for him to re-evaluate the situation. However, due to the circumstances of that flight, he thought it would have been much more effective to hear the FO clearly state his concern, the problem, and propose the solution. While it would have been easy for the debriefer to explain this, the lesson learned was more meaningful coming from another crewmember and specifically the fellow crewmember of the flight.

As the T-list is completed, I check to see if every lesson learned is written as closely as possible to the way that the crewmember stated it. I also confirm that all statements stand alone and bring out the details that would bring about an improvement, or reinforce what was done well.

I use this time at the end of the debriefing to clear up any confusion as to what each individual was thinking or meant by their actions or statements during the flight. I ask the SME for questions I may have missed, or to add any comments germane to the discussion. If there seems to be any unresolved personal or operational conflict, this is the time I attempt to resolve it. To resolve personal conflict, the debriefer needs to emphasize the issues that the conflicting parties agreed upon and the positive things they said about each other. Then, if the discussion progresses towards whether or not they would fly with each other again, these things can be mentioned while expressing the company's expectation that they *would* fly with each other. The expectation to resolve personal conflict should be realistic. It is the awkward tension of past personal conflict that makes it so difficult to resolve. After conducting over 140 Mediated Debriefs, I can say that no one left without shaking hands and expressing they would fly together again. While I have been told that is an impressive record, I am not so naïve as to be overly excited. Many of those did not involve personal conflict. Of the ones that did, I doubt many of the crewmembers involved would consider each other friends, but they definitely left the debriefing with at least the tools and the attitude to be able to fly together again with the miscommunication resolved and the expectations of each expressed and heard.

*Recap.* At this point the debriefer reads back what the crewmembers have said they would improve next time, relating what they have said to what is being taught in the standards department of the company, to reinforce the learning. The debriefer finishes the recap by reading the *Well* side of the T-list. As tense as the Mediated Debriefs can be, I have never had any complaints by ending on the most positive note. This is a very simple and effective method of reinforcing the learning by making crewmembers feel more comfortable. However, this insight escapes many aviation instructors. When many instructors critique, they speak in *general* terms of what the crewmembers did well, and usually become *specific* only during the areas needing improvement. Yet, in fact, the impressive safety record of U. S. airlines is the direct result of the highly manifest skills of crewmembers. On every flight crews perform many tasks extremely well. Instructors who do not acknowledge this explicitly lose credibility with crews. Fair and accurate representation of all aspects of the crewmembers' performance is crucial.

The debriefing is concluded by asking the crew if they have any remaining questions for anyone in the room. After addressing any questions, the debriefer thanks the crew for their time, emphasizing how much everyone in the room has learned and the value of the lessons learned for the crewforce.

## Dissemination of Lessons Learned

De-identified summaries of Mediated Debriefs can be a valuable source of information for the entire crewforce and can be used as thought-provoking training material. The debriefer can prepare a summary from the meeting notes, starting with a descriptive synopsis of the events of the incident flight. Material from the *Well* and *Improve* sections from the T-list should be included, reworded only enough for readers to understand. By staying as close as possible to the crew's own words, the debriefer can summarize the discussion in a way with which other flight personnel can identify.

The summaries must be carefully de-identified, both to avoid embarrassing the crew and to reduce vulnerability to litigation. Simply deleting names may not be sufficient de-identification. The debriefer strikes a balance between summarizing events in sufficient detail for lessons to be learned and not providing enough information that someone might infer the identity of the crew.

## An Example: Near Controlled Flight Into Terrain (CFIT) Incident

The following is an actual write-up from a Mediated Debrief involving an airline crew that became confused with their approach clearance, descended dangerously close to terrain, and had to execute a GPWS escape maneuver. The three-person crew was on a night flight in visual meteorological conditions (VMC).

> This was the first time this crew had flown together. The CA had about 2 years pilot-in-command (PIC) in aircraft type, the FO had less than 100 hours in his seat position, and the SO had only about 6 months line experience. Both the FO and SO said the CA did an excellent job of briefing the crew before departure. For example, the CA briefed the SO to take initiative concerning the fuel dump if an airborne problem developed without needing to ask. [Most airlines require the SO to first request permission of the CA to start dumping. This CA's briefing could save valuable seconds in an emergency.] He also briefed the new FO to fly first leg based on weather, weight of aircraft, terrain and FO's experience. Enroute, the crew briefed the terrain and the expected DME arc approach.
>
> When handed off to a new controller, who was difficult to understand, they were cleared to descend directly to the destination over terrain instead of staying on their flight-planned route. The flight was given a clearance for the VOR/DME *transition* to the runway. The CA, expecting the VOR/DME *arc* approach, read back that he understood cleared for the DME *arc* to intercept the final. The controller failed to respond and correct the readback error. The FO then asked if they should begin the turn to intercept the arc. The CA said he guessed they should and the FO began to turn to intercept the arc. After the aircraft rolled out from the turn, they received a GPWS alert. The FO stated that since he could see the lights of the city in front and below him, he was surprised by the GPWS and hesitated. He later stated that after the GPWS alert he lost situational awareness for four minutes. The CA stated he expected the FO to start the GPWS escape procedure immediately but since he did not, the CA reached to push the power forward. The CA stated he thought this situation was real and was very concerned. During the GPWS maneuver the SO called out 'Radar altimeter!' which was distracting to the CA.
>
> The controller asked for their heading and told them they were supposed to have gone directly to the VOR. The FO started to lower the nose and the SO cautioned them to pull back the power. The CA took control of the aircraft as he picked up the runway visually then turned toward the field. Once stabilized the CA gave control of the aircraft to the FO who made an uneventful landing.
>
> Upon arrival the crew debriefed the flight. Thanks to this debrief, the uneventful return flight was flown by a well-coordinated crew, using effective

communication. The CA requested the Mediated Debrief to gain closure and disseminate lessons learned.

As per the procedures of the Mediated Debrief, the SO gave his summary, followed by the FO and finally the CA (see Figure 6.2 for the actual T-list from this Mediated Debrief). When the FO began to speak, he explained he at no time felt they were in any danger when he heard the GPWS. By the end of the Mediated Debrief, he exclaimed that since they were at 13,000 feet above sea level, and the ground was less than 300 feet below, he now realized how close they had come to CFIT. He also remarked that a friend of his had died flying military aircraft in the same area due to less than accurate terrain data. The CA was thankful to put closure on the incident by discussing with his crew all of the concerns they identified in the debriefing. He stated he would ensure that never again would he continue flying with ambiguous clearances from Air Traffic Control (ATC). He would also maintain greater situational awareness with regard to aircraft position and terrain. Also, he would follow instrument procedures for arrivals and Standard Terminal Arrival Routes (STARs).

The CA was quick to point out that his recent recurrent training was extremely useful for dealing with the emergency they experienced. The simulator training included GPWS maneuvers, which was realistic and relevant. The ground school also included lessons learned from past Mediated Debriefs. One of those included a near CFIT in which a crew had descended into high terrain on a night visual approach.

## Conclusion

The Mediated Debrief is a powerful approach to helping crews learn from their mistakes in a constructive manner. It also generates knowledge that can be shared with the entire workforce to enhance safety. Airlines have used this knowledge to improve operating procedures, flight operations manuals, CRM training, and technical training. The feedback from participants in the debriefings and from those who have read the incident summaries has been extremely positive.

Beyond aviation, I have been asked to implement the program in European hospitals, working with surgeons and nurses to review simulated and actual operations. I am currently working with a city school district that has implemented a system in which teachers work in teams to deal better with the contemporary challenges of education. In principle, this method can be adapted to any situation involving team performance.

| | WELL | IMPROVE |
|---|---|---|
| **SO** | • Knowledgeable about SOP's.<br>• Effective assertion – called for power reduction after the GPWS maneuver.<br>• Excellent job monitoring other crewmembers.<br>• Exemplary performance especially for his experience level.<br>• Concise in his communications. | • Will maintain greater SA of terrain in approach area.<br>• Will not call out radar altitude during GPWS maneuver. |
| **FO** | • Gives attention to detail.<br>• Effective communicator and effective team skills.<br>• Nice landing. | • Will execute GPWS escape maneuver, phase one, immediately. |
| **CA** | • Effective team formation/management (i.e. established atmosphere of active participation).<br>• Very thorough – delegated workload.<br>• Performed GPWS escape maneuver and exercised authority as per FAR 91.3a.<br>• Nice landing. | • Clarify all clearances or communication with ATC to eliminate any ambiguity whatsoever.<br>• Fly the standard instrument procedures for arrival and approach.<br>• Will maintain higher SA of position with regard to destination and terrain. |

**Figure 6.2 T-List With Lessons Learned**

## Notes

1 As this person may come from the Standards/Evaluation Department, there is the unwanted possibility of this individual jumping into the stereotypical role of the 'accusing' check airman. The debriefer precludes this by stating ground rules, individual roles, and procedures during the introduction.

2 This is not always possible, for various reasons. For example, an incident may come to the attention of the safety department only after the FAA has notified the crew of a past problem.

3 However, crewmembers should not be asked to tell each other what they should do to improve.

4 Inexperienced debriefers frequently fail to address the debriefings the crews conduct at the end of flights, perhaps because line debriefings are often neglected in airline operations (see chapter 5). In Mediated Debriefs, crews often reveal that the flight legs after the one in which a problem occurred were even more dangerous because of the resulting breakdown in crew communication. CAs could resolve these situations by conducting a debriefing immediately after the problem flight. In one actual line situation, an effective CA stopped the aircraft at the hold short line before takeoff to address his crew. They had just inadvertently crossed an active runway without a clearance. The CA took a minute to gather the attention of his stunned crew and tell them he took responsibility for the error, but they had a job to complete, and they would debrief the incident after completing the flight. In this way the CA helped the crew focus on the task at hand without preoccupation with the unsettling event.

# 7 Facilitating a Critical Incident Stress Debriefing

MIMI TOMPKINS

## Introduction

The aftermath of an aircraft accident is often replayed on television news over and over for days and weeks. The surviving passengers tell their stories. The family members of those who did not survive grieve while millions watch. Crewmembers and their families usually do everything possible to remain out of the spotlight.

Until just a few years ago airline crewmembers involved in accidents were expected to get on with their lives and jobs as soon as possible, without any psychological support. Airline employees also impacted by the accident would continue working despite lacking an outlet for their own grief.

In the early to mid-1990s pilot and flight attendant unions, sometimes working together with management, developed programs to provide psychological support to crewmembers, other company employees, and their families after an accident or incident. These programs involve the management of critical incidents or accidents using peer volunteers under the leadership of mental health professionals. Peer volunteers are trained to facilitate debriefings with the crewmembers or employees directly involved in or effected by an incident or accident.

The purpose of the critical incident response programs is to mitigate the long-term stress reactions of crewmembers. Without any support or counseling an accident or incident may effect work performance, the crewmember's career, relationships, financial fitness, or health.

In this chapter, I will discuss critical incident stress debriefings (CISD) and the role of the peer/facilitator.

## Definition of a Critical Incident

A critical incident is defined as an event that has the potential for people to experience significant physical, cognitive, emotional and behavioral reactions immediately after the incident or days, weeks, months and even sometimes years later. 'It is an event which is considered generally outside the range of ordinary human experience' (Everly and Mitchell, 1996, p.62).

Specific events listed in the Air Line Pilot Association's Critical Incident Response Program Guide (ALPA, 1998) that require intervention by peers and mental health professionals are:

1. an aircraft accident;
2. an aircraft incident that results in injury to persons or damage to equipment;
3. an on-the-job incident in which exists the real or perceived threat of death or serious injury to a crewmember or passenger, i.e., bomb threat or near mid-air collision;
4. an incident requiring passenger evacuation; or
5. participation in aircraft accident investigation activities (ALPA, 1998, p. iii).

Less extreme incidents may also traumatize crewmembers. For example, in one incident a flight attendant cut her finger while opening a wine bottle. The severed artery spurted blood over several rows of passengers. This life-threatening situation was described in an incident report as a 'cut finger', yet, several of the crewmembers experienced stress reactions.

For crewmembers, the accident/incident does not actually have to occur for stress reactions to result. Donald Hudson, MD, ALPA Aeromedical Advisor, related an incident in which the Captain of a major air carrier 'almost' landed with the gear up. A crewmember noticed the gear was not down on very short final and the Captain instigated an immediate and successful go-around. The aircraft incurred no damage and no one was injured, yet, the Captain was diagnosed with a severe and long-term stress disorder.[1]

Critical incident stress management has also been implemented within the airline industry for such diverse events as: (1) turbulence resulting in injuries to flight attendants; (2) a crewmember critically injured by ground equipment on the ramp; (3) catastrophic engine failure resulting in passenger deaths; (4) death or serious injury of crewmembers' children; (5) severe wind shear; (6) death of a ramp agent struck by a turning propeller; (7) pilots listening to the cockpit voice tape as part of an accident investigation; and (8) hotel layover events such as witnessing shootings, robberies, death of crewmembers in their sleep, and hotel van accidents.

## Critical Incident Stress Management

The process of critical incident stress management (CISM) is 'an integrated system of interventions which are designed to prevent and or mitigate the adverse psychological reactions' that may follow an accident or incident (Everly & Mitchell, 1996). Typically, the process consists of three parts: pre-incident education, defusings, and debriefings. I will discuss each of these in the following sections.

Research on the effectiveness of critical incident stress management programs is sparse. Formal studies comparing the outcome of these programs with other interventions or with no intervention have not been performed, perhaps in part because it seems unethical to withhold support to some personnel. However, other methods of evaluating program effectiveness are possible and empirical information would be useful.

Aside from formal studies, we can get some sense of the value of these programs from the reports of individuals who have benefited from them. During 1998, over 300 pilots participated in CISM defusings and debriefings after critical incidents.[2] The feedback received from many of these participants was that the program was helpful in understanding the impact and resolving issues that came up for them about the incident and how it was handled.

Personal discussions reveal that one value of the program is the peer-to-peer interaction. Crewmembers have often commented that they felt supported because someone from their peer group cared enough to make contact, really listened to them relate their experience, and followed up with resources when needed.

Also, peer volunteers become a bridge to link crewmembers who need professional support with mental health professionals. In training, peer volunteers meet and get to know mental health professionals who attend basic critical incident stress management with them. They are then able to recommend these professionals to the crewmembers who may need additional counseling.

### *Pre-Incident Education*

Understanding that it is normal for a person, even a trained crewmember, to experience stress reactions following an abnormal or emergency flight or ground event is the first and most important step in developing critical incident stress management programs. A critical incident stress defusing or debriefing in the aftermath of a critical incident is much easier to facilitate if the crewmembers and their management supervisors have been educated prior to the event. With pre-incident education, standardized training, clear program guidelines and protocols, everyone involved knows what to expect. Several airlines have included modules on critical incident stress management in crew resource management or ground training for crewmembers.[3]

*Defusings*

Before a debriefing is conducted, a defusing will be done. A defusing is a short 20-30 minute facilitated discussion best conducted within 8 to 24 hours following a critical incident. It is called a defusing to distinguish it from the more formal critical incident stress debriefing (CISD) that always includes a mental health professional.

Even before this defusing is conducted peer volunteers contact the crew as soon as possible after the event in order to begin minimizing stress reactions by providing for physical and logistical needs. The crewmembers will be moved away from the scene of the incident or accident to a quiet, secure, and private location. They will be shielded from the media and authorities who are a required part of the incident/accident process. Medical care will be offered and the need for such care assessed by someone other than the crewmembers themselves. Water and easy-to-digest, nutritious food will be made available. Going for a brisk walk also helps reduce stress reactions.

After the crewmembers' basic needs are met and the situation has calmed down, a group or individual defusing will be held with everyone who participated directly in the incident or accident. If is it not possible to conduct a group defusing before everyone leaves the scene, individual telephone defusings will be conducted after each person gets home or into a hotel room.

This intervention has a three-fold objective: (1) it prevents crewmembers from feeling isolated or abnormal for experiencing stress reactions, (2) it provides support from other crewmembers, and (3) it provides professional resources if additional support is needed.

During the defusing the peer asks the crewmembers what happened and encourages them to talk without interruption. Next he or she asks if they recognize any physical or psychological impact from the event or how it was handled. At the end of the defusing, the peer educates the crewmembers in a short lecture format about the effects of stress and how to cope with it. The peer also provides handouts about normal stress reactions and what can be done to minimize these reactions.

Permission may be obtained to contact the crewmembers' significant other or another family member. It is important that family members also understand that stress reactions are 'normal reactions to an abnormal event'

(Everly & Mitchell, 1996), even for trained aircrew professionals. Often another handout about stress reactions and ways to handle the stress will be mailed to the crewmember's home, addressed to both the crewmember and significant other.

Follow-up calls are made over the next few days to make sure there were no delayed stress reactions. Depending on the stress reactions of the crewmembers, follow-up will be continued for several weeks. If the event involved the death or serious injury of a passenger or crewmember, a more formal critical incident stress debriefing involving a trained mental health professional will be arranged.

## The Critical Incident Stress Debriefing [4]

The critical incident stress debriefing (CISD) is a facilitated psychological debriefing developed by Jeffrey T. Mitchell as a 'direct, action-oriented crisis intervention process designed to prevent or mitigate traumatic stress subsequent to a traumatic event' (Everly and Mitchell, 1996). There are many other debriefing models used following traumatic events,[5] however, the airline industry is currently using the CISD as taught by the International Critical Incident Stress Foundation located in Ellicott City, Maryland.

The debriefing allows participants to talk about their experience, thoughts, and feelings in a supportive group. The process of talking, in some cases, may be cathartic and it can help participants understand what happened and how they reacted. It also provides a way for participants to support and console each other.

A peer and a mental health professional facilitate the debriefing, as described later. These individuals use facilitation techniques to elicit participation, which may at first be difficult for the crewmembers, and to keep discussion focused on the crewmembers' experience, reactions, and concerns.

As crewmembers speak aloud about an event, they begin to make more sense of what happened. As they listen to the other crewmembers, the event takes shape more clearly. It may take 24 to 48 hours following a traumatic event to begin remembering details in the order they actually happened. For example, in an incident involving an off-course navigation error, each crewmember remembered events in a different order.

Logistically, arranging a critical incident stress debriefing involves juggling the diverse schedules of crewmembers or other company employees, often living in several different cities and working varied shifts. If a crew is scheduled to participate in a company operational debriefing, FAA meeting, or NTSB interview, it is convenient to arrange the debriefing for the same day, before or after the other meeting. If the event involved the death of company employees, it will be necessary to wait until after any funeral(s).

To improve the effectiveness of operational debriefings, one airline schedules a critical incident stress debriefing with crewmembers prior to the operational debriefing. The stress debriefing usually helps relieve some of the normal tension and confusion that a crew may be experiencing after a critical event. They can then be more objective in discussing the technical and procedural details of what happened when they participate in the operational debriefing with the company.[6]

*Preparation for the Debriefing*

As with any facilitated group interaction, preparation for the critical incident stress debriefing is essential. Pre-incident education concerning critical incident stress management with crewmembers makes it easier to arrange and facilitate this debriefing. If the crew has no concept of 'psychological first aid' or the reason for stress management, it may be a challenge getting all involved in the incident to agree to meet together with a mental health professional.

Management may also be suspicious of the purpose of a debriefing. Pre-incident education of management supervisors helps them understand the importance of CISM and elicits their cooperation in arranging debriefings.

Often supervisors want to be part of the debriefing, either because they care about their employees or because they want to feel in control. However, a basic rule of stress debriefings is that only the people who were directly involved in the event may be present in the debriefing, often excluding supervisors. No one who has disciplinary responsibility may be present unless they were directly part of the accident or incident, and in these exceptional cases everyone else participating in the CISD must give permission for them to participate. Normally, supervisors should be provided their own debriefing with other CISD trained management peer personnel.

The peers facilitating the CISD arrive before the scheduled meeting time to familiarize themselves with the event they are going to be addressing and to arrange the room. The room used should be quiet, private, and comfortable. Chairs are arranged in a circle so each participant can see the others. Peers and the mental health professional preselect their chairs in order to be able to see each other easily. They space themselves evenly among the group members and may use their eyes or slight hand motions to communicate with each other.

As the participants arrive for the CISD, the team members circulate and talk with them. Casual conversation helps reduce tension and allows the participants to get to know the team members.

The facilitators meet alone outside the room for a few minutes after the group participants are encouraged to take care of their personal needs and instructed where to find the restrooms. A checklist of the seven steps of the debriefing may be reviewed and handed out to use in order that nothing is missed. Also, the peers and mental health professional plan last minute strategy. Here the roles of each peer are prearranged.

The team facilitation leader is usually, but not always, the mental health professional. Peers are assigned the roles of doorkeeper and co-leader. The doorkeeper will (1) keep out people who should not be attending the debriefing, (2) welcome participants who may be late, and (3) attempt to persuade anyone who leaves the room while the debriefing is in progress to return back to the group. If the person does not want to return, the peer stays with that person and continues the debriefing one-to-one, if possible. The co-leader listens closely to the team leader and fills in anything that is missed. He or she also 'teaches' the stress management information during the teaching phase near the end.

### *Seven Steps of a Critical Incident Stress Debriefing*

The critical incident stress debriefing has seven steps: introduction, fact, thought, reaction, symptom, teaching, and re-entry (Everly & Mitchell, 1996, p.101). The order of these seven steps is important; however, it is not always possible to follow them precisely. Emotions like grief and sadness may be expressed as soon as people are given an opportunity to talk. It is important for the facilitator to let the participants talk even if what is being said is not relevant to that particular step.

The *introduction* phase of the critical incident stress debriefing outlines the purpose of the meeting, the phases of the debriefing, and ground rules. Usually one of the peers begins by discussing the guidelines and ground rules. This discussion helps decrease resistance, reduce anxiety, and motivate the group to participate freely.

The main rules or guidelines for a critical incident stress debriefing that must be agreed upon by all participants include: (1) everyone present must have been directly involved in the event (no one is allowed to sit in to observe); (2) confidentiality must be acknowledged by everyone present; (3) it is okay to describe someone else's reactions but not okay to give them meaning such as 'She was terrified.'; (4) no one has to speak if he or she does not want to; (5) questions may be asked at anytime; (6) other housekeeping rules, such as turning off pagers and phones; and (7) circumstances under which individuals may or may not leave the room.

In the *fact* phase, the mental health professional asks the participants to introduce themselves, state their role in the incident, and share what they experienced. Each person is given an opportunity to speak in order around the circle. Even though the participants usually already know each other, these easy to answer questions help them get comfortable with the process of talking in the group. Also, this may clear up any misunderstanding about each person's role in the incident.

In the *thought* phase, participants are asked about their first thoughts during the incident. For example, if they saw someone get injured, they might be asked, 'What were your thoughts when you saw that person on the floor?' Or in the case of a turbulence incident, 'What went through your mind when you realized turbulence was much worse than normal?' Asking about the first thought puts each participant mentally back at the scene. Often people are surprised at their first reaction to trauma. Talking about these thoughts and hearing others express their feelings helps the participants accept that their thoughts and feelings were okay, or normal.

In the *reaction* phase participants are asked a question such as, 'What was the worst part of this incident for you?' Individuals often form 'snap-shot' visual memories of traumatic events and participants may be asked, 'If you could erase any of these memories, what would you choose to delete?' During this phase participants speak when they want to; they do not each talk in turn around the circle.

Talking about the worst part may be very emotional. Peers must be able to sit and listen to a range of emotions, many of them very sad. Sometimes individuals do not want to express their emotions in front of others and will just shake their heads indicating that they do not want to speak. In this case the facilitator just goes on to the next person without pressuring the individual to speak. Sometimes there is only silence for many seconds.

In the *symptom* phase the mental health professional or peer begins to bring the participants back into a less emotional state. Participants are asked about their own stress reactions during the event, immediately after, and in the present. The facilitators may have to give examples of different types of stress reactions and how they may effect people cognitively, emotionally, physically, and behaviorally. Each person will have to think about his or her reactions and will usually be less emotional.

In the *teaching* phase the group is asked what they know they can do to relieve their stress. The co-leader steps out of the facilitator role and supplements what the participants already know with a tutorial on how diet, exercise, alcohol, caffeine, nicotine, and other factors affect stress. He or she also facilitates this phase by reviewing stress reactions already mentioned by the participants in earlier phases of the debriefing and relating these reactions to what can be done to dissipate them. All of the peers eventually participate in the teaching phase by adding stress management strategies that the co-leader may have left out.

In the *re-entry* phase, each person is asked, 'Is there anything positive that has been learned from this event?' or 'What good has come out of this event?' This phase will also help bring the participants out of the emotional stage and should help them think and feel more positively about their experiences.

In this last phase, the participants are also asked to introduce any new material they wish to discuss, review anything already discussed, ask any questions they might still have, and discuss anything that would help them bring closure to the debriefing. The facilitators will answer any questions, reassure and inform as needed, identify any unspoken thoughts or unacknowledged emotions that are observed but not discussed previously, provide appropriate handouts, and summarize (Everly & Mitchell, 1996, p. 112).

At the end of the debriefing, each facilitator makes a short summary to encourage participants to think about their reactions to the event as normal. Facilitators often conclude by expressing their appreciation for the participation of the crewmembers.

### *After the Debriefing*

As soon as the summary is finished, one of the team facilitators invites participants to help themselves to the refreshments that were set out before the debriefing. Drinking healthy liquids and eating a light snack such as fruit may relieve anxiety and is a way for the participants to share and join together. Most group participants choose to mingle before leaving for home.

The facilitators walk around and make personal contact with each participant, paying special attention to any who were particularly distressed or unusually quiet. This provides a way to reach out to those who may need additional or individual support. Some people (less that 1 to 3%) may need referral for individual therapy (Everly & Mitchell, 1996, p. 114).

The facilitators also arrange to follow-up with each participant with a phone call or brief visit. If appropriate, the mental health professional may meet with an individual participant in a one-on-one session in another room immediately after the debriefing. The CISD team remains until all of the participants leave.

After the other participants leave, the peers and mental health professional meet to discuss and self-critique the CISD. During this meeting the peers arrange follow-up with specific participants and assign any remaining tasks for the incident/accident response. The facilitators discuss the CISD process, what went well and what could be improved.

The facilitators also discuss how they were affected by what they experienced in the debriefing. If the event that precipitated the debriefing was profoundly distressing, a critical incident stress debriefing will be arranged for the peers and mental health professional.

In addition to the crew and peer debriefing, modified critical incident stress debriefings may be scheduled for all company employees following a major aircraft disaster. These debriefings keep the employees informed about the event, educate them about stress reactions, give them an opportunity to express their thoughts and feelings, and make sure they know where they may turn for additional help if needed.

Follow-up is one of the most important aspects of critical incident stress management and without follow-up the effectiveness of the debriefing may be reduced. Telephone calls, individual meetings, referrals for therapy, additional meetings with the group, family support, and other contacts may be necessary.

For a major aviation accident, follow-up is recommended after any widespread media coverage, investigation disclosures, NTSB hearings, the anniversary of the accident, and publication of the NTSB finding. Follow-up may be needed for many months and even periodically over several years. For example, Donald Hudson, M.D., of the ALPA Aeromedical Advisory Office, observed that his office begins to hear from crewmembers suffering the long-term effects of stress around 18 months after the event.[7]

## Debriefing Facilitation Techniques

During the debriefing, the peers and mental health professional must model good listening skills.[8] The peers should be very alert to verbal and non-verbal cues such as tone of voice and obvious changes in body language. Mirroring the body language and facial expressions of the participants, paraphrasing, restating, and clarifying to check if what was heard was the intended message are also important.

Some listening skills that aid in the process of the CISD involve 'not doing' something: for example, remaining neutral by not displaying a preference for solutions participants suggest, not trying to 'fix' the participants' problems, not denying the feelings heard ('you shouldn't feel that way'), not making decisions for the participants, not speaking while letting the participants display their feelings, and not displaying shock at what they say.

The peers and mental health professional do as little talking as possible, rather using questions and other facilitation techniques to elicit and focus participation. Typically the questions used in a debriefing start out as closed questions, designed to illicit information and get the participants comfortable talking, and then progress to open ended questions requiring longer responses and deeper participation by the crewmembers. These open-ended questions are the heart of the facilitation process.

Facilitators help the participants reflect on and voice what they are thinking and feeling. One of the most important skills for facilitators is voicing what they are thinking and feeling to the group (Schwarz, 1994). The facilitators' sincere personal reactions model the behavior they want to see and hear in the participants. For example, the facilitator feels sad and angry after listening to a crewmember describe how he felt holding the hand of a ramp agent who had just been killed by the aircraft's propeller. If the crew is not expressing their thoughts or feelings, the facilitator may say how she feels and why. This validates that it is normal to have a range of feelings as a result of the situation.

## What Makes a Good Facilitator?

Peers are recruited from all backgrounds within an airline. Many have little or no previous experience in a facilitation process; however, they typically do very well, if given appropriate training in facilitation skills. The most important skill for the facilitator is good listening. As simple as it may seem, good listening is the most challenging skill for non-professional facilitators. In-flight pilots are rewarded for their excellent problem-solving abilities. Yet orientation toward problem solving – vital in complex, time sensitive, safety related tasks – often gets in the way when pilots first attempt to use good listening skills.

After participating in many debriefings, I have come to realize that certain facilitation skills are critical. The peer facilitator who enters the debriefing with a 'big picture' of what he or she would like to see happen tends to be successful. The CISD big picture consists of the crewmembers expressing what the event was like for them, what they were thinking and are thinking about the event, and helping them understand that their feelings or thoughts are normal. Successful facilitators use the previously mentioned 'seven phases' as a road map for accomplishing this.

A successful facilitator recognizes when to step out of the facilitator role and become an 'expert' by providing information only he or she has and which the participants need in order to continue the process. This is specifically done in the teaching phase of the debriefing but may be needed in any of the phases.

A good facilitator does not tell his or her own 'war' stories. It may be appropriate to mention they have been through their own accident or incident, but the focus must remain on the participants and this event. A simple statement such as 'I have seen a lot of adversity in my life' may be enough to make the point.

Facilitators should not become over invested in their own interpretation of what happened. For example, a good facilitator will not assume to know what was the worst part of an incident for a crewmember. Good facilitators employ their excellent listening skills with the patience to allow each participant to say in their own words or way what is bothering them.

Despite all the facilitator training, listening skills, and techniques, participants rarely remember any precise thing a facilitator says in a defusing or a debriefing. The process the participant goes through in the debriefing is what counts. Each participant will remember strong emotions, vividly described images, and their own personal experience. People taking the time to care, to listen, and to extend a hand by listening when others may be feeling isolated and alone is the heart of defusing and debriefing.

The ultimate goal is to facilitate the normal coping skills of crewmembers, so they may get back to work as quickly as possible. Thanks to the critical incident stress management facilitation process, trained and dedicated volunteers have accelerated the recovery process for crewmembers following incidents and accidents.

## Notes

1 Donald Hudson, MD, Air Line Pilots Association Aeromedical Advisor at the ALPA Air Safety Forum, Washington, DC, 1991.

2 Air Line Pilots Association Human Performance Committee report to the ALPA President and Executive Council, May 1999.

3 As of this publication, Air Wisconsin, Alaska Airlines, Aloha Airlines, Trans World Airways, and US Airways have formally introduced CISM.

4 Information about CISD was taken from training and books by George T. Everly, Jr., Ph.D. and Jeffrey T. Mitchell, Ph.D., co-founders of the International Critical Incident Stress Foundation, Inc. and the author's experience since 1994 as a peer volunteer for crewmembers.

5 The American Red Cross utilizes a different debriefing with surviving passengers and their families after aircraft accidents.

6 Captain Bill Morin, Alaska Airlines, ALPA CIRP Chairperson discussed the benefits of CISDs at an ALPA CIRP Chairperson meeting April 20,1999 in Herndon, VA.

7 Donald Hudson, MD, Air Line Pilots Association Aeromedical Advisor at the ALPA Air Safety Forum, Washington, DC, 1991, and in a telephone conversation with the author September 1999.

8 McDonnell, Jobe, and Dismukes (this volume) discuss facilitation techniques in detail.

## References

Air Line Pilots Association. (1998). *Critical incident response program guide.* Virginia: Author.

Everly, G., & Mitchell, J. (1996). *Critical incident stress debriefing CISD*, 2nd ed. revised. Maryland: Chevron Publishing Corporation.

Schwartz, R. (1994). *The skilled facilitator*. San Francisco: Jossey-Bass.

# Appendix

## DEBRIEFING ASSESSMENT BATTERY

**INSTRUCTOR PROFILE**

The Instructor Profile is a summary of the strategies and techniques IP's use to assist crews in conducting their own debriefings while giving direction and focus as necessary. The two main goals of the debriefing are to 1) get the crew to perform an in-depth analysis of the situation that confronted them, how they understood and managed the situation, the outcome, and ways to improve, and 2) get the crew to participate in a proactive, rather than reactive, manner in which they initiate discussion and elaborate beyond the minimal. These goals are based on the assumption that active participation by the crew will result in a higher level of learning and increased likelihood of transfer to the line.

**Directions:**

Use the scale below to rate the instructors on each of the following elements, then total the scores to get the overall rating for each category.

| Poor | Marginal | Needs Improvement | Adequate | Good | Very Good | Outstanding |
|---|---|---|---|---|---|---|
| 1 | 2 | 3 | 4 | 5 | 6 | 7 |

**Introduction**

One purpose of the introduction is to let the crew know that participation and self-evaluation are expected of them, and why it is important.

_______Makes clear that his role is guide/facilitator and that crew should do most of the talking

_______Clearly conveys that crew should take an active role, initiating discussion rather than just responding to him

_______Clearly conveys that he wants crew to dig deep, critically analyzing the LOFT and their performance

_______Gives a persuasive rationale for the crew to participate actively and make their own analysis

_______**Overall rating of Introduction**

**Questions**

The purpose of asking questions is to get the crew to participate, focus the discussion on important topics, and enlist the crew in discussing the topics in depth.

_______Asks an appropriate number of questions to get crew talking & lead them to issues

_______Avoids answering for the crew when they do not respond immediately or correctly and uses a pattern of questioning that keeps the focus on the crew

_______Uses probing and follow-up questions to get crew to analyze in depth and to go beyond yes/no and brief factual answers

_______Uses questioning techniques to encourage interaction and sharing of perspectives among crew members

_______**Overall rating of Questions**

**Encouragement**

Encouragement refers to the degree to which the instructor encourages and enables the crew to actively and deeply participate in the debriefing.

_______Conveys sense of interest in crew views and works to get them to do most of the talking

_______Encourages continued discussion through active listening, strategic pauses, avoiding disruptive interruptions, and/or following up on crew-initiated topics

_______Encourages all members to participate fully, drawing out quiet members if necessary

_______Refrains from giving long soliloquies or giving his own analysis before crew has fully analyzed

_______**Overall rating of Encouragement**

**Focus on Crew Analysis and Evaluation**

The goal of the debriefing session is to get the crew to evaluate and analyze their own CRM performance so they will learn more deeply and can gain practice in debriefing themselves, a skill they can then begin to use on the line.

_______Encourages crew to analyze along CRM dimensions the situation that confronted them, what they did to manage the situation, and why they did it

_______Encourages crew to evaluate their performance and/or ways they might improve

_______Encourages crew to explore CRM issues and how they specifically affect LOFT performance and line operations

_______Encourages crew to analyze issues, factors, and outcomes in depth, going beyond simply describing what happened and what they did

_______**Overall rating of Focus on Crew Analysis & Evaluation**

**Use of Videos**

One stated purpose of showing videotaped segments of the LOFT is to enable the crew members to see how they performed from an objective viewpoint so they can better evaluate their performance. More realistically, perhaps, the video reminds the crew of the situation, aiding their memory and providing a focus for discussion.

_______Shows an appropriate number of videos of appropriate duration to illustrate/introduce topics

_______Uses video equipment efficiently: is able to find desired segment without wasting time and pauses the video if substantial talk begins while playing

_______Consistently discusses video segments, using them as a springboard for discussion of specific topics

_______Has a point to make and uses the video to make that point.

_______**Overall rating of Use of Videos**

**CREW PROFILE**

The crew profile measures the degree and depth of participation by the crew.

**Directions:**

Use the scale below to rate the crew on each of the following elements, then total the scores to get the overall rating for each category

| Poor | Marginal | Needs Improvement | Adequate | Good | Very Good | Outstanding |
|---|---|---|---|---|---|---|
| 1 | 2 | 3 | 4 | 5 | 6 | 7 |

**Crew Analysis and Evaluation**

Crew analysis and evaluation refers to the depth to which the crew members analyze the LOFT situation and evaluate their performance.

_______Analyze along CRM dimensions the situation that confronted them, what they did to manage the situation, and why they did it
_______Evaluate their performance and ways they might improve
_______Explore CRM issues and how they affect LOFT performance and line operations
_______Analyze issues, factors, and outcomes in depth, going beyond simply describing what happened and what they did
_______**Overall rating of Crew Analysis & Evaluation**

**Depth of Crew Activity**

Activity refers to how actively, versus passively, and deeply the crew participates in and initiates discussion.

_______Go beyond minimal responses to IP questions
_______Participate deeply and thoughtfully
_______Initiate dialogue rather than just responding to questions, and/or interact with each other rather than only with the IP
_______Behave in a predominantly proactive rather than reactive manner, being actively involved rather than just passing through the training
_______**Overall rating of Depth of Crew Activity**

# Index